Michael Kreutz

The Renaissance of the Levant

Judaism, Christianity, and Islam – Tension, Transmission, Transformation

Edited by Patrice Brodeur, Alexandra Cuffel, Assaad Elias Kattan, and Georges Tamer

Volume 13

Michael Kreutz

The Renaissance of the Levant

Arabic and Greek Discourses of Reform in the Age of Nationalism

DE GRUYTER

ISBN 978-3-11-064356-5
e-ISBN (PDF) 978-3-11-063400-6
e-ISBN (EPUB) 978-3-11-063134-0
ISSN 2196-405X

Library of Congress Control Number: 2019931215

Bibliografic information published by the Deutsche Nationalbibliothek
The Deutsche Nationalbibliothek lists this publication in the Deutsche Nationalbibliografie; detailed bibliografic data are available on the Internet at http://dnb.dnb.de.

This volume is text- and page-identical with the hardback published in 2019.
Printing and binding: CPI books GmbH, Leck

www.degruyter.com

Contents

Acknowledgments — VII

Introduction — 1

1 New Ideas, Crumbling Orders — 11
1.1 Cultural Entanglements — **11**
1.2 Enlightenment in Perspective — **17**
1.3 Theology and Science — **26**
1.4 Agents of Reform — **36**
1.5 Napoleon in Egypt — **49**

2 The Mediterranean Dawn — 55
2.1 The Greek Rise to Independence — **55**
2.2 The End of Eastern Decline — **63**
2.3 A Feminist Revolution — **77**
2.4 Towards a New Humanism — **84**
2.5 From Turkey to Japan — **99**

3 Civilizations Drifting Apart — 111
3.1 The Past and the Future — **111**
3.2 Mapping Out New Orders — **129**
3.3 Eastern anti-Westernism — **141**
3.4 After Disillusionment — **152**

4 Concluding Thoughts — 157

Timeline — 160

Sources — 165

Literature — 167

Index — 183

Acknowledgments

I would like to express my warmest thanks to Assaad Kattan (Münster) for his wonderful intellectual support for my project under the original title "Beyond the centers: Religion and Enlightenment by the example of Greek and Arab Christianity in Southeast Europe and the Middle East (*Jenseits der Zentren: Religion und Aufklärung am Beispiel von Islam, griechischem und arabischem Christentum in Südosteuropa und im Vorderen Orient*); to the DFG for funding this book; to the Cluster of Excellence at the University of Münster, i.e. the working platform *Transcultural Entanglements*, for providing a conducive environment; to the members of my project group *Transfer between world religions* for their feedback; to Katharina Linnemann (Münster) for reviewing and improving the manuscript; and to David West (Münster) for careful scrutiny of my English. Last but not least, I would like to thank my parents and friends most heartily for their invaluable and strong backing while I was writing this book.

https://doi.org/10.1515/9783110634006-001

Introduction

Since the Mediterranean connects cultures, Mediterranean studies have by definition an intercultural focus. This is a vast field and requires a command of different languages not necessarily related to each other. Throughout the modern era, the Ottoman Empire has had a lasting impact on the cultures and societies of the Southern and Eastern Mediterranean. It is for this reason that it makes sense to investigate both Greek and Arabic sources – two essential languages in that area that are connected by the fact that Orthodox Christians have written in both of them. This book will shed some light on the significance of ideas in the political transitions of their time, and on how the proponents of these transitions often became so overwhelmed by the events that they helped trigger adjustments to their own ideas.

The book is therefore both a survey of intellectual history from the perspective of mainly Greek and Arabic speakers. Greek and Arabic challenge attempts to modernizing culture in their own special way since they are languages being associated with the Bible and the Quran. founding texts of two great civilizations. Being languages of erudition Greek and Arabic used to connect different peoples over a vast stretch of territory which is another challenge to political reforms oriented towards an order of nation states where languages no matter what historical baggage they carry have to be forged into national languages. On a final note, the discourses in Greek and Arabic reflect the provinces of the Ottoman Empire and it will be interesting to see their differences and commonalities. The discourse which is I examine here is pivoting around religion but with a secular outlook.

I.

Intellectual history has become a neglected field of research and a nineteenth century remnant[1] dismissed by historians and political scientists alike for reasons not entirely clear. In Germany it was not properly restored after the Second World War,[2] but has received more academic attention during the last ten years

1 Isaiah Berlin and Ramin Jahanbegloo, *Den Ideen die Stimme zurückgeben: Eine intellektuelle Biographie in Gesprächen* (Frankfurt/Main: Fischer, 1994), 118; Wilhelm Schmid, *Die Geburt der Philosophie im Garten der Lüste* (Frankfurt/Main: Suhrkamp, [1987] 2000), 23.

2 Wilhelm Bleek, *Geschichte der Politikwissenschaft in Deutschland* (Munich: Beck, 2001), 16–9; Paul Nolte, *Transatlantische Ambivalenzen: Studien zur Sozial- und Ideengeschichte des 18. bis*

https://doi.org/10.1515/9783110634006-002

or so. In the English-speaking world intellectual history is largely associated with the Cambridge School and its proponents, Quentin Skinner and JGA Pocock. Skinner's approach was akin to that of Leo Strauss, who defined the task of any reader as being to understand the author as he or she understood him or herself.[3] However, Skinner was criticized by Hans-Georg Gadamer, who argued that any attempt to reveal the original intention of the author of a source was bound to fail.[4] The Cambridge School earned its stripes when it managed to rewrite parts of the history of English political thought based on its analysis of patterns of thought typical of certain strata of society at a certain period of time.[5] However, its equation of the limits of language with the limits of human action was unconvincing, since language can adapt to the most diverse political situations, which means that the individual human does not surrender to concepts that lie beyond his or her influence. There was a similar discussion about the Frankfurt School when Jürgen Habermas made language an ineluctable element of human understanding, something that was criticized by CS Pierce, who argued that there existed nothing in this world that warranted its own prerequisites.[6]

As the political scientist Hannah Arendt argued, however, intellectual history is key to understanding the present, since political ideas function as reference parameters for political actors to decide and judge in times of uncertainty.[7] Much of what intellectual history is able to unearth is the struggle for liberty and its double sense, both positive and negative, which becomes manifest in political independence as well as in the rule of law. We often find that both ideas are

---

20. Jahrhunderts (Berlin: De Gruyter, 2014): 391–5; cf. Michael Kreutz, "Ideengeschichte als Blindstelle historischer Forschung in Deutschland," in *michaelkreutz.net*, Aug. 16, 2017, URL=http://www.michaelkreutz.net/2017/ideengeschichte-als-leerstelle-der-heutigen-forschung/ [Aug. 24, 2018].

3 Klaus Oehler, *Blicke aus dem Philosophenturm: Eine Rückschau* (Hildesheim, Zurich and New York: Olms, 2007), 187; cf. John GA Pocock, "Sprache und ihre Implikationen: Die Wende in der Erforschung des politischen Denkens," in *Die Cambridge School der politischen Ideengeschichte*, ed. Martin Mulsow and Andreas Mahler (Berlin: Suhrkamp, 2010): 88–126, here 113.; Quentin Skinner, *Visionen des Politischen* (Frankfurt/Main: Suhrkamp, 2009), 81.

4 David Harlan, "Der Stand der Geistesgeschichte und die Wiederkehr der Literatur," in *Die Cambridge School der politischen Ideengeschichte*, ed. Martin Mulsow and Andreas Mahler (Berlin: Suhrkamp, 2010): 155–202, here 161, 165–7, cf. ibid. 189.

5 Ibid., 168–9.

6 Klaus Oehler, *Blicke aus dem Philosophenturm*, op. cit., 382–4.

7 Grit Straßenberger, "Hannah Arendt, Michael Walzer und Martha Craven Nussbaum," in *Politische Ideengeschichte im 20. Jahrhundert: Konzepte und Kritik*, ed. Harald Bluhm and Jürgen Gebhardt (Baden-Baden: Nomos, 2006): 155–80, here 165.

closely intertwined; and, while they sometimes complete each other, they might also fall into opposing camps.

"The very name" of liberty is in any case forgotten sometimes, and falls into oblivion when other values and ideals become more important. The Eastern Mediterranean of the modern era provides numerous examples of how progressive ideas based on the idea of liberty gave way to more authoritarian traits in society, and even paved the way for them when the very modern state that they had helped to create fell prey to an ethnic nationalism that was focused on expansion and hegemony in the region. This is a kind of "dialectic of enlightenment," and it is one of the goals of this book to make clear how the intertwining of progressive, universalist values on the one side, and ethnic nationalist currents on the other, can help understand the current politics of identity in the Eastern Mediterranean.

II.

This fact has been obfuscated by the Orientalism debate. According to Edward Said (2003), "to be a European in the Orient," one must "see and know the Orient as a domain ruled over by Europe." Orientalism, which is the system of European or Western knowledge about the Orient, thus becomes synonymous with European domination of the Orient."[8] He defines Orientalism as "a style of thought based upon an ontological and epistemological distinction made between 'the Orient' and (most of the time) 'the Occident'." Therefore, writers of all kind and in all fields have accepted the basic distinction between East and West as "the starting point for elaborate theories, epics, novels, social descriptions, and political accounts" in terms of the Middle East and Islam.[9] This has all supposedly become part of a Western hegemony over the Middle East and itself a Western cultural practice that perpetuates the "disparity in power between the West and non-West."[10]

Yet, Said and the postcolonial school perpetuate this binary view, which does not reflect the realities on the ground: Greece and all of Southeast Europe do not easily fit into this binary pattern, if they fit at all. Said does not recognize that the predominantly Christian-Orthodox part of the European continent has historically never been part of the Occident, and is usually not viewed nowadays

8 Edward Said, *Orientalism* (London et al.: Penguin, 2003), 197.
9 Ibid., 2–3.
10 Said, *Culture and Imperialism*, 1994, 230.

as being part of the Orient, either. Besides, it was not the "West" that dominated the Middle East, but almost exclusively Great Britain and France, which Said well knew. And both Britain and France often saw themselves as being superior not only to the Middle East, but also even to many cultures in mainland Europe. Moreover, when they were secular, the same forces that aimed to overcome French and British imperialism also aimed to overcome Ottoman imperialism. In so doing, they also harnessed the British and the French powers to achieve their own ends.

Although the ancient Mediterranean has long since disappeared (no matter whether we follow Henri Pirenne's thesis or not), the modern Eastern Mediterranean shares the common burden of an Ottoman legacy that resists any binary thinking in terms of West and non-West. Toner (2013) is very much to the point when he argues that "[t]he relationship between East and West ... was always a complex dialectical phenomenon."[11] He agrees that European thinkers had found something of a mirror in the "Orient," i.e., in mainly the provinces of the Ottoman Empire.[12] But, given the complexity of West-East relations and the many facets of mutual perception, we cannot but conclude that "[t]here is no simple postcolonial model that can be applied retrospectively." Moreover, most of the English discourse on the Turks in the seventeenth century was driven less by colonialism than by economic success, although both goals may have overlapped.[13]

The dichotomy between East and West became a permanent feature of Western perception, with Europe and Asia being placed in opposite positions. Nevertheless, the notion that the Orient was somehow connected to Europe and constituted the cradle of European civilization was not yet over.[14] During the Enlightenment, Western intellectual attention to non-European cultures may

---

11 Jerry Toner, *Homer's Turk: How Classics Shaped Ideas of the Middle East* (Cambridge/Mass. and London: Harvard University Press, 2013), 14–5.

12 Ibid., 113.

13 Ibid., 75–6; cf. Michael Marx, "Europa, Islam und Koran: Zu einigen Elementen in der gegenwärtigen gesellschaftlichen Debatte," in *Gehört der Islam zu Deutschland? Fakten und Analysen zu einem Meinungsstreit*, ed. Klaus Spenlen (Düsseldorf: Düsseldorf University Press, 2013): 61–98; Hartmut Fähndrich, "Orientalismus und Orientalismus: Überlegungen zu Edward Said, Michel Foucault und westlichen 'Islamstudien'," in *Gegenwart als Geschichte: Islamwissenschaftliche Studien: Fritz Steppat zum fünfundsechzigsten Geburtstag*, ed. Axel Havemann and Baber Johansen (Leiden et al.: Brill, 1988): 178–86; Ulrike Freitag, "The Critique of Orientalism," in *Companion to Historiography*, ed. M Bentley (London and New York: Routledge, 1997): 620–38; Edward Said, "Orientalism Reconsidered," *Race and Class* 27, Vol. II (1985): 1–15.

14 Jürgen Osterhammel, *Die Entzauberung Asiens: Europa und die Asiatischen Reiche im 18. Jahrhundert* (Munich: Beck, 2010), 53, 55.

have been selective and Eurocentric, as Kohl (1987) has argued, but this was a result of their critical view of absolutist rule at home, which contrasted with supposed moderate rule abroad. Such approaches to non-European cultures were not necessarily driven by any sense of superiority.[15] The idea that Europe owed much to Asia, and especially in terms of Christianity, would weaken during the nineteenth century[16] due to the destruction by the Ottomans of ancient sites in the Balkans and in Asia Minor.[17] With the decline of Ottoman power, the European tendency to emphasize cultural differences weakened again.[18]

There has been criticism of Said's claims even within postcolonial studies. Although well aware that for a British nineteenth-century traveller the Orient was identical to India,[19] Said does not arrive at the conclusion that, as Devji (2016) has argued, India also provided the blueprint for Britain's understanding of Islam. This, though, contradicts Said's claim that Western perceptions of Islam derive from Western cultural resources.[20] Contrary to Said's thought, it is the pan-Islamic idea in which Muslims and non-Muslim scholars had found a common topic independent that was untouched by any Orientalist fantasy. The pan-Islamic idea was brought into being by the Treaty of Küçük Kaynarca that was signed in 1774 by Russia and the Ottoman Empire, with this Treaty allowing Russia to protect the sultan's Orthodox subjects while the sultan was granted religious authority over the Muslim population outside his empire. This was a historical novelty.[21] Besides that, Britain's foremost enemy had been and continued to be Tsarist Russia – not the Ottomans or the Persians.[22]

None of this seems to be of relevance to Said. And, although he provides some strong arguments to show the racist character, especially of scholarship in the nineteenth century, he goes too far when he argues that the "modern Ori-

---

15 Karl-Heinz Kohl, *Abwehr und Verlangen: Zur Geschichte der Ethnologie* (Frankfurt/Main and New York: Campus, 1987), 126.
16 Osterhammel, *Die Entzauberung Asiens*, op. cit., 52.
17 Ibid., 47–9.
18 Toner, *Homer's Turk*, op. cit., 104.
19 Edward Said, *Orientalism*. op. cit., 169.
20 Faisal Devji, "Islam and British Imperial Thought," in *Islam and the European Empires*, ed. David Motadel (Oxford: Oxford University Press, 2016): 254–68, here 257. Said: "... modern Orientalism derives from secularizing elements in eighteenth-century European culture." Edward Said, *Orientalism*, op. cit. 120. "Thus whatever good or bad values were imputed to the Orient appeared to be functions of some highly specialized Western interest in the Orient." Ibid., 206.
21 Devji, "Islam and British Imperial Thought," op. cit., 258, cf. Annemarie Schimmel, *Die Zeichen Gottes: Die religiöse Welt des Islams* (Munich: Beck, 1995), 233.
22 Devji, "Islam and British Imperial Thought," op. cit., 261; cf. Efraim Karsh, *Islamic Imperialism: A History* (New Haven and London: Yale University Press, 2006), 95–8.

entalist" does not "stand apart" from the Orient "objectively" since his attitude is marred by "the absence of sympathy." This means that "[h]is Orient is not the Orient as it is, but the Orient as it has been Orientalized."[23] From this angle, not only does sympathy become a major requirement for understanding the Middle East, but any proper understanding has to be proven against "the Orient as it is." In other words, Said is completely unaware that sources require interpretation, and that "the Orient as it is" is itself an ideological formula.[24] Thus, for Said, the modern Orientalist is a complete slave to error, and Said's rage against him frustrates any critical approach towards the Islamic Middle East in particular, and towards the Islamic world in general. The creation of a "new literature," something that Said advocates, a literature that aims to provide an authentic depiction of the native culture of the Orient, a culture "not pristine and pre-historical," but instead "deriving from the deprivations of the present," is another such ideological formula, since it reduces scholarship to a simple reproduction of local self-conceptions.[25] This is corroborated in Said's 1994 book *Culture and Imperialism*, where he lashes out against academics who are "opposed to native Arab or Islamic nationalism."[26]

Said opposes capitalism, too, describing it in a neo-Marxist manner as a system of subjugation that is "commanded at the top by the handful of leading industrial countries."[27] This is not convincing since, as Fukuyama (2014) has pointed out, large parts of the Middle East lack a middle class that would foster a stable democracy.[28] Accordingly, Acemoglu and Robinson argue in *Why Nations Fail* (2012), that "it was the expansion and consolidation of the Ottoman Empire, and it is the institutional legacy of this empire that keeps the Middle East poor

---

23 Edward Said, *Orientalism*, op. cit., 104. "Yet the Orientalist remained outside the Orient ..." Ibid., 222. Later on, Said backed out and said: "This is not to denigrate the accomplishments of many Western scholars, historians, artists, philosophers, musicians, and missionaries, whose corporate and individual efforts in making known the world beyond Europe are a stunning achievement." Edward Said, *Culture and Imperialism* (London: Vintage, 1994), 235, cf. ibid. 342.
24 Said speaks of Orientalism as a "school of interpretation" and an "academic tradition," but one that is "all aggression, activity, judgment, will-to-truth, and knowledge" so that a scholarship which Said would not reject must be one that merely reproduces what local peoples believe of themselves, ibid. 203–4.
25 Ibid.,272.
26 Ibid., 315.
27 Ibid., 272, 320.
28 Francis Fukuyama, *Political Order and Political Decay: From the French Revolution to the Present* (London: Profile Books, 2014), 435; cf. 431.

today."[29] They do not deny that Western imperialism has had a devastating impact on colonialized nations,[30] but argue that that does not explain why some remain poor. Some do, they argue, because those in power pursue poor governance.[31]

A counter example to the poverty of the Middle East with its Ottoman legacy is, to quote Fukuyama again, East Asia, whose major states "were constructed around bureaucratic cores that owe more to their own historical experience than to anything imported from the West."[32] Local cultural resources can therefore explain a great deal of the current state of the Eastern Mediterranean. Said, however, is entirely focused on cultural exchange and, for him, there is no doubt that any such exchange between Europe (i.e. Western Europe) and the Middle East can only take place at the expense of the latter. In other words, what the Occident takes from the Orient is only for the benefit of the former. From this angle, the colonized peoples are necessarily on the side of the losers, irrespective of whether the cultural exchange was imposed or voluntary. The net effect of cultural exchange between two partners who are conscious of their inequality cannot be anything but negative: "someone loses, someone gains."[33] In the same book, however, Said also argues that "the history of all cultures is the history of cultural borrowings" which he describes as a universal and global phenomenon. By arguing that culture "is never just a matter of ownership [...], but rather of appropriations, common experiences, and interdependencies"[34] he rebuts his own conviction that in terms of cultural exchange "someone loses, someone gains."

As Khawaja (2008) has pointed out, Said "feels free to offer large-scale generalizations" about all kinds of phenomena – with the sole exception of Islam, "which appears not to be governed by any rule." Khawaja also argues that Said rejects any idea of a true or false Islam on the one hand, while, on the other, praising certain writers for understanding Islam correctly and reprimanding others for misunderstanding Islam. To make matters even more confusing, "Orientalist" research stands accused of being too abstract, but also of being too con-

29 Daron Acemoglu and James A. Robinson, *Why Nations Fail: The Origins of Power, Prosperity, and Poverty* (New York: Crown Business, 2012), 56, cf. ibid. 61.

30 "[...] European expansion into the Atlantic fueled the rise of inclusive institutions in Britain." This in turn has caused an underdevelopment in the colonized areas. Ibid., 250.

31 Ibid., 68.

32 Fukuyama, *Political Order*, op. cit., 537.

33 Said, *Culture and Imperialism*, op. cit., 235.

34 Ibid., 261.

crete in its view of Islam, which means that, as Khawaja puts it, "[t]o be consistently inconsistent about consistency is a defeat in itself."[35]

Despite all these problems, Said's binary view has become mainstream. What is even worse, though, is that there is an alarming trend in the humanities that sees a privileging of origin and social background over expertise and knowledge of the sources. A recent example is Pankaj Mishra's 2012 book *From the Ruins of Empire*,[36] which, adhering to the same binary view, sees Europe as a deeply destructive force. The latest such attempt at constructing a binarity is Bauer's (2011) description of the Islamic world as a superior civilization shaped by a culture of ambiguity and eventually weakened by European values. I have given a short account of the most significant flaws in that argument in my 2016 book *Zwischen Religion und Politik.*[37] Bauer's argument is the result of seeing the Islamic world within the framework of a binary view, one that sees East and West as two contesting entities. However, Bauer fosters yet another binary view when he depicts the Islamic world as a somewhat superior entity that did not go into decline before the nineteenth century when it tried to adopt European ideals and manners. Against this view I understand both civilizations as overlapping entities with the Eastern Mediterranean being a transition zone and I shall argue that the predominantly Orthodox nations of Southeast Europe still today have more in common with Arab countries than with Western Europe.

III.

Though complementary, Mediterranean studies (more specifically, Levantine studies) and the adjacent Balkan and Middle Eastern studies remain strangers.[38] In my books *Arabic Humanism in the Modern Era* (*Arabischer Humanismus in der Neuzeit*, 2007) and *The End of the Levantine Age* (*Das Ende des levantinischen Zeitalters*, 2013), I have tried to depict the Eastern Mediterranean in terms of cultur-

---

35 Irfan Khawaja, "Essentialism, Consistency and Islam: A Critique of Edward Said's *Orientalism*," in *Postcolonial Theory and the Arab-Israel Conflict*, ed. Philip Carl Salzman and Donna Robinson Divine (London and New York: Routledge, 2008): 12–36, here 26–7, 29.

36 Pankaj Mishra, *From the Ruins of Empire: The Intellectuals Who Remade Asia* (New York: Picador, 2012).

37 Thomas Bauer, *Die Kultur der Ambiguität: Eine andere Geschichte des Islams* (Berlin: Verlag der Weltreligionen, 2011). Michael Kreutz, *Zwischen Religion und Politik: Die verschlungenen Pfade der Moderne* (Bochum: Verlag Michael Kreutz, 2016), 79–102.

38 One of the few exceptions is Karl Kaser, *The Balkans and the Near East: Introduction to a Shared History* (Vienna and Berlin: Lit, 2011) which however covers a far longer period of time and focuses less on modern developments.

al entanglement at a time when ended what I have coined the "levantine age," i.e. the old imperial order with its mixed populations.[39] It makes perfect sense to study the Arab East in the context of the Eastern Mediterranean and thus of countries such as Greece and Albania. One of the first scholars to pave the way for this kind of approach was Papoulia (2003), who dealt with the transformation of the Balkans from an imperial system to a system of nation-states. He did so by analyzing the layers of Ottoman-Islamic rule that are crucial for the understanding of the transformation process.[40] The imperial heritage finds an echo in what has recently been termed "phantom borders," which define memory spaces that are rooted primarily in the nineteenth century.[41]

The transformation of the imperial order into one of nation states was driven and accompanied by discourses of cultural revival, nationalism, and enlightenment. Though entangled, these led to different results in the Eastern Mediterranean. I will focus in the following on both enlightenment and nationalism as two manifestations of cultural revivalism rooted in the same intellectual discourse of the time, and explore their interrelationship. Although they shared a common strategic goal, i.e., establishing a modern, non-despotic nation-state, they marked the beginning of very different developments. All of this took place against the background of an intellectual challenge that found its symbol in Europe (as an idea rather than a continent). Europe was the place where modernity had become manifest, and secular intellectuals aspired to becoming part of it. After the Second World War, however, Mediterranean culture was marginalized

---

39 "The external military threat of the Turks, which had put Europe in a state of fear through the 1520's and 1530's, diminished, and the internal climate became more favorable for favorable attitudes to Turkish culture. [...] In 1536, under the French ambassador Jean de la Forest, an expedition was dispatched to research the Ottoman Empire and Islam, as well as to collect Oriental and classical manuscripts. The scientific leadership of this expedition was taken over by the Orientalist and mathematician Postel. In the text that Postel wrote at the conclusion of his research expedition, he confronted the Christian world with an extremely positive view of Turkish culture." Dietrich Klein, "Hugo Grotius' position on Islam as described in *De Veritate Religionis Christianae, Liber VI*, in *Socinianism and Arminianism: Antitrinitarians, Calvinists and Cultural Exchange in Seventeenth-Century Europe*, ed. Martin Mulsow and Jan Rohls (Leiden und Boston: Brill, 2005): 149–73, here 153; see also Clarence D. Rouillard, *The Turk in French History, Thought, and Literature (1520–1660)* (Paris: Boivin, 1941), 105–6 and 207–12, and Rudolf Franz Merkel, "Der Islam im Wandel abendländischen Verstehens," in *Studi e Materiali di Storia delle Religioni* 13 (1937): 68–101, here 83.

40 Basiliki Papoulia, *Από την αυτοκρατορία στο εθνικό κράτος* (Thessaloniki and Athens: Ekdoseis Banias, 2003), *passim*.

41 Dan Diner, "Zweierlei Osten: Europa zwischen Westen, Byzanz und Islam," in *Das Europa der Religionen*, ed. Otto Kallscheuer (Frankfurt/Main: S. Fischer, 1996): 97–113, here 99.

within the discourse of modernity.[42] This has changed in recent years because the European Union seeks to foster good relations with the countries on the other side of the Mediterranean. Yet, this has found a rather faint echo in the humanities; and not only in Germany, where no established Mediterranean studies exists, and where what is labelled as such often does not stand up to closer examination. For example, research on Christian convents in Spain, or on Ottoman rule in Hungary, or on the modern transformation of the Balkans do not come under the umbrella of Mediterranean studies, since they fail to identify the Mediterranean as a common cultural space. This has some relevance for Mediterranean studies, but the emphasis here is on "some," since it is not part of Mediterranean studies in a strict sense, which would involve the Mediterranean as the connecting element between cultures.

In my 2013 book *Das Ende des levantinischen Zeitalters*, I attempted to weave the ideas and the events of the modern Eastern Mediterranean into one continuous narrative. Weaving a common narrative in which the Mediterranean connects rather than separates was a personal challenge not only because of the complexity of the matter, but also because it meant setting foot on new ground. This book develops my previous one, and accentuates further the commonalities and differences between the processes of transformation that led to the modern system of nation-states, while providing at the same time a modern view of Europe from its periphery.

This book comprises three main chapters. The first focuses on both the ideas and the events that challenged the old imperial order in the Eastern Mediterranean, beginning in the early nineteenth century. This led to events that would dismantle the Ottoman Empire, events that were accompanied by a political discourse conducted in different languages across the shores. This discourse in the late nineteenth and early twentieth centuries aimed to foster a cultural revival which was seen as a prerequisite for the making of modern nation-states, and was a discourse that was largely secular and progressive, and that focused on cultural and social reform. However, as I will show in my last chapter, war and nationalist radicalization made room for a more authoritarian kind of thinking that has brought trouble to many Mediterranean polities today. Of special interest here are the sectarian frameworks in this transformation process and their ongoing legacy in contemporary politics.

42 "Mittelmeerkultur ist im Zeichen transatlantischer Weltaneignung immer mehr ins Abseits des Diskurses der Moderne geraten." Georg Stauth and Marcus Otto, *Méditerranée: Skizzen zu Mittelmeer, Islam und die Theorie der Moderne* (Berlin: Kadmos, 2008), 17.

1 New Ideas, Crumbling Orders

1.1 Cultural Entanglements

The Enlightenment has had more than one center and has never been restricted to Europe. Although its geographical centers were England, Scotland and France, it is France in particular that became associated with Enlightenment ideas due to the propagation of these ideas during the eighteenth century.[1] The Enlightenment and the idea of Europe were intertwined, with the effect that Europe was strongly identified in the Middle East with modernity. Southeast Europe adopted a more ambivalent attitude than the Middle East, however, since it believed that it was the Latin west that had attempted to subdue the Orthodox east for its own ends. In any case, "Europe," as Burke (1980) has put it, "is not so much a place as an idea."[2] As an evolutionary concept, "Europe" is much more than a geographical term; and, since Europe lacks a distinct boundary to the East, the definition of "Europe" has been the subject of debate.

The term "Occident" long denoted the Western part of Europe and excluded the predominantly Slavic and Orthodox east of the continent. The defining elements were the Catholic faith and the realm of the Latin culture of learning. During the Renaissance, humanists preferred the notion of "Europe" to that of the "Occident," since the former evoked memories of the Greek and Roman antiquity, while "Occident" did not. The Occident became even more strongly associated with Latin Christianity when Luther coined the term "Orient" to denote the realm of Islam in opposition to Christianity.[3] This did not mean that the term "Europe" became less popular. On the contrary, as Burke has pointed out:

---

1 Richard Clogg, "The ‚Dhidhaskalia Patriki' (1798): An Orthodox Reaction to French Revolutionary Propaganda," in *Middle Eastern Studies*, Vol. 5, No. 2 (May, 1969): 87–115, here 87.

2 The confusion of West and Europe is a common mistake, cf. Peter Burke, "Did Europe exist before 1700?," in *History of European Ideas*, Vol. 1/1 (1980): 21–29, here 21.

3 Michael Kreutz, *Das Ende des levantinischen Zeitalters: Europa und die Östliche Mittelmeerwelt, 1821–1939* (Hamburg: Kovac, 2013), 59–63; Heinz Gollwitzer, *Europabild und Europagedanke: Beiträge zur deutschen Geistesgeschichte des 18. und 19. Jahrhunderts* (Munich: Beck, 1964), 33, 39; Josef Matl, *Südslawische Studien* (Munich: R. Oldenbourg, 1965), 147–9; Thomas Kaufmann, "Kontinuitäten und Transformationen im okzidentalen Islambild des 15. und 16. Jahrhunderts," in *Judaism, Christianity, and Islam in the Course of History: Exchange and Conflicts*, ed. Lothar Gall and Dietmar Willoweit (Munich: R. Oldenbourg, 2011): 287–306, here 288, 296; Almut Höfert, *Den Feind Beschreiben: ‚Türkengefahr' und europäisches Wissen über das Osmanische Reich, 1450–1600* (Frankfurt/Main and New York: Campus, 2003), 62. Helmut Wilsdorf, "Georgius Agricola, die 'jüngeren Griechen' und das Morgenland," in *Griechenland – Byzanz – Europa: ein Studienband*, ed. Joachim Herrmann, Helga Köpstein and Reimar Müller

https://doi.org/10.1515/9783110634006-003

> Thus for nearly two thousand years, from the fifth century BC to the fifteenth century AD, the term "Europe" was in sporadic use without carrying very much weight, without meaning very much to many people. From the later fifteenth century, however, it came to be taken rather more seriously ... In the sixteenth century, most leading writers used the term ... In the seventeenth century, references multiply still further.[4]

According to this line of argumentation, the term "Europe" was often used in secular contexts and was in a relationship of tension with the notion of "Christendom,"[5] although it rests on the same pillar (namely, the Roman heritage). The Roman heritage is twofold, however, and those cultures that grew on the remnants of the Eastern Roman Empire still to some extent remain distinct in their political culture.[6] Huntington (1996) has argued that the divide between Eastern and Western Europe echoes the major traits of Christianity and can be traced back to the fourth century and the Roman period. This corresponds in the Balkans to the historical border between the Austro-Hungarian and the Ottoman Empire. According to Huntington, Europe ends "where Western Christianity ends and Islam and Orthodoxy begin."[7] Therefore, and if Haffner ([1985] 2001) and Huntington are correct, modern Southeast Europe should be studied in the context of the Middle East. The latter is itself a disputed term that overlaps partly with the "Levant," which, as can be seen in the travelogues of the time, originally, and especially in the eighteenth and nineteenth centuries, denoted the whole of the Eastern Mediterranean. Like the term "Occident," it implied an ideological delimitation regarding the "Turkish threat."[8]

"Levant" also used to be associated with the major port cities of the Eastern Mediterranean, well-known for their cosmopolitan societies and permissive attitudes. Both positive and negative reflections on this can be found in modern lit-

(Amsterdam: Gieben, 1988): 215–24, here 220; cf. Johannes Irmscher, "Über den Morgenland-Begriff," in *Byzantino-Sicula* II. *Miscellanea di scritti in memoria di G. Rossi Taibbi* (Palermo: Istituto siciliano di studi bizantini e neoellenici, 1975): 295–300, *passim*; Andrea Polaschegg, *Der andere Orientalismus: Regeln deutsch-morgenländischer Imagination im 19. Jahrhundert* (Berlin: De Gruyter, 2003), 64–5.

4 Burke, "Did Europe exist before 1700?," op. cit., 23–4.

5 Ibid., 27.

6 Sebastian Haffner, "Das Nachleben Roms," in idem, *Historische Variationen* (Stuttgart and Munich: DVA, [1985] 2001): 31–8, here 32, 34.

7 Samuel P. Huntington, *The Clash of Civilizations and the Remaking of World Order* (New York: Foreign Affairs, 1996), 158.

8 Kreutz, *Das Ende des levantinischen Zeitalters*, op. cit., 20–2.

erature.[9] Italian humanism began to show an interest in the fifteenth century in studying Eastern Roman literature, which was followed in the sixteenth century by the editing of vernacular texts. German Protestantism turned its attention at the same time to the Greeks under the Ottoman yoke, attention that was manifested in the 1584 *Turcograecia*, edited by Martin Crusius (1526–1607). Crusius had also been in contact with the Venice-based humanist Maximos Margounios as well as with the Greek theologian Theodosios Zygomalas, counsellor to the Greek patriarch Meletios Pegas. Pegas was himself acquainted with Latin culture, since he had studied in Padua.[10] Protestant contacts with Greek theologians intensified during the sixteenth century, when Philipp Melanchthon began correspondence with Corfu-based Antonios Eparchos and the patriarchate of Constantinople, and created a network of intellectual exchange between Constantinople, Heidelberg and Tübingen.[11]

The same holds true for Western contacts with the Islamic realm. Many European scholars were fascinated during the final period of the Renaissance (a "time of awakening to travel," Osterhammel 2010) by Persia, which had embarked on a program of modernization initiated by Shāh ʿAbbās I (r. 1588–1626), who attempted to open his country carefully to Western European influence.[12] When ʿAbbās managed in 1623 to break the Portuguese maritime trade monopoly, he was helped to do so by Britain's East India Company, which had been founded in 1600. The good ties with a Western power added to a romanticized image of non-European societies, the latter, with their apparent embodiment of justice and equality, contrasting sharply with the absolutist regimes of Western Europe.[13]

The Treaty of Sitvatorok was signed by Austria and the Ottoman Empire at roughly the same time (in 1606). This was the first time that the sultan had recognized Europeans as negotiating partners on equal terms. A major shift occurred in 1648 with the Treaty of Westphalia, when the Ottoman Empire, although not a signatory state, was integrated into a system of imperial states with clearly

---

9 Nathalie Clayer, "Der Balkan, Europa und der Islam," in *Wieser Enzyklopädie des Europäischen Ostens*, Vol. 11: *Europa und die Grenzen im Kopf*, ed. Karl Kaser, Dagmar Gramshammer-Kohl and Robert Pichler (Klagenfurt, Vienna and Ljubljana: Wieser, 2003): 303–30.

10 Gerhard Podskalsky, *Griechische Theologie in der Zeit der Türkenherrschaft (1453–1821): Die Orthodoxie im Spannungsfeld der nachreformatorischen Konfessionen des Westens* (Munich: Beck, 1988), 85–6, 102–3, 128.

11 Georg Ludwig von Maurer, *Das griechische Volk in öffentlicher, kirchlicher und privatrechtlicher Beziehung vor und nach dem Freiheitskampfe bis zum 31. Juli 1834*, Erster Band [1st vol.] (Heidelberg: Mohr, 1835), 18.

12 Jürgen Osterhammel, *Die Entzauberung Asiens*, op. cit., 103.

13 Ibid.; Karl-Heinz Kohl, *Abwehr und Verlangen*, *op. cit.*, 126.

defined borders.[14] At the same time, Western views of Islam became more pluralized.[15] Barthélemi d'Herbelot, who held a chair in Oriental languages in Paris, published his *Bibliothèque orientale* in 1697, which was an encyclopedia that dealt with the peoples of the Orient in more than 8,000 entries. He was unable to finish the task of examining the huge corpus of Oriental manuscripts that the encyclopedia was based on, but his work was continued by his friend Antoine Galland. The latter shared d'Herbelot's view of Islam as something that was not in opposition to Europe and European culture, but rather as something that belonged to a common space where the civilizations of the antiquity met and the classical heritage was transmitted to the contemporary world.[16]

In terms of foreign politics, the Ottoman Empire became part of a system of imperial powers controlling the Mediterranean. The architecture of peace had been oriented since its beginnings towards achieving a balance of power and a relationship of cooperation irrespective of religion. This is why Edmund Burke in 1765 called the Ottoman Empire "a great power of Europe," a description that reflected a popular stance at the time.[17] When the Russian empress Catherine the Great annexed the Crimea in 1783, the Western public did not ally themselves with Russia, but instead condemned Catherine's actions.[18] However, the idea that the Ottoman Empire was in a way part of Europe did not remain uncontested. Burke withdrew his statement in 1791, and began to regard the Ottoman Empire as part of Asian civilization. And Herder, who had always propagated a rather non-exclusivist definition of Europe, considered the Turks on the European continent as foreigners.[19] The Ottoman Empire had also as a

14 Von Maurer, *Erster Band* [1st vol.], op. cit., 24; Ralf Elger, "Selbstdarstellungen aus Bilâd ash-Shâm: Überlegungen zur Innovation in der arabischen autobiographischen Literatur im 16. und 17. Jahrhundert," in *Eigene und fremde Frühe Neuzeiten: Genese und Geltung eines Epochenbegriffs*, ed. Renate Dürr, Gisela Engel and Johannes Süßmann (Munich: R. Oldenbourg, 2003 = *Historische Zeitschrift*, Beiheft 35): 123–37, here 127–8; cf. Edgar Hösch, *Geschichte der Balkanländer: Von der Frühzeit bis zur Gegenwart* (Munich: Beck, [4th] 2002), 109.

15 Klein, "Hugo Grotius' position," op. cit., 154, 157–8; cf. Jan Brugman and Frank Schröder, *Arabic Studies in the Netherlands* (Leiden: Brill, 1979), 4–5, and Johann Fück, *Die arabischen Studien in Europa bis in den Anfang des 20. Jahrhunderts* (Leipzig: Harrassowitz, 1995), 63.

16 Osterhammel, *Die Entzauberung Asiens*, op. cit., 56–7.

17 Ibid. 47.

18 Robert K. Massie, *Catherine the Great: Portrait of a Woman* (New York: Random House, 2011), 67; Efraim and Inari Karsh, *Empires of the Sand: The Struggle for Mastery in the Middle East, 1789–1923* (Cambridge/Mass. and London: Harvard University Press, [1999] 2001), 4, 12–5, 16–7.

19 Osterhammel, *Die Entzauberung Asiens*, op. cit., 48. "Am wenigsten kann also unsre Europäische Cultur das Maas allgemeiner Menschengüte und Menschenwerthes sein; sie ist kein oder ein falscher Maasstab. Europäische Cultur ist ein abgezogener Begriff, ein Name. Wo existirt sie ganz? bei welchem Volk? in welchen Zeiten?" Johann Gottfried Herder, in *Herders Werke*,

matter of fact been Europeanized by being placed within the framework of the Troian War, which connected it to Greek history.[20] France had always enjoyed good relations with the Ottomans, since both were engaged in the slave trade. But fascination for the Ottomans ran deeper and culminated in an expedition in 1536 to collect Islamic manuscripts. This expedition was championed by the French writer Guillaume Postel (d. 1581), who adopted quite a positive attitude towards the Ottomans. This was still an exception, since relations between the Western European states and the Ottoman Empire were tense. However, Ottoman penetration into Europe seemed to have ceased by the second half of the sixteenth century, and Western Europe found a new interest in Ottoman culture.[21]

With the beginnings of English and French colonial expansion in the Middle East, we can find by the end of the eighteenth century, as Bernard Lewis has reminded us, a number of Western European grammars on Middle Eastern languages. According to Lewis, there were seventy on Arabic, ten on Persian, and fifteen on Ottoman Turkish. There were also ten textbooks for Arabic, four for Persian, and seven for Ottoman. Although Ottoman was both the *lingua franca* and the administrative language of the Ottoman Empire, grammars and textbooks on Arabic far outnumbered those for Ottoman-Turkish. This was because Arabic, as the language of the Quran, was considered to have the same importance as Latin, Greek and Biblical Hebrew. Lewis' remarks are directed subtly against Edward Said's claim that (Western) Europe never saw Arabic and Islam as being on a par with their own culture.[22]

Yet, modern Greek seemed to be excluded from European culture. This, at least, was the accusation leveled at Karl Krumbacher by the Greek linguist Georgios Hatzidakis. The Byzantine scholar Krumbacher had written in his 1903 book *The Problem of Modern Greek Literary Language* (*Das Problem der neugriechischen Schriftsprache*) of a Greco-Slavonic part of Europe. For Hatzidakis, this was nothing but an unjustified marginalization and exclusion of Greek culture.

ed. Heinrich Meyer, Hans Lambel and Eugen Kühnemann, Vol. 5, Pt. 2: "Briefe zu [sic!] Beförderung der Humanität" (Stuttgart: Union Deutsche Verlagsgesellschaft, 1889): 542.

20 Toner, *Homer's Turk*, op. cit., 78.

21 Klein, "Hugo Grotius' position," op. cit., 153–4; Marion Kuntz, *Guillaume Postel: Prophet of the Restitution of All Things: His Life and Thought* (The Hague: M. Nijhoff, 1981), 4; Wilhelm Schmidt-Biggemann, "Political Theology in Renaissance Christian Kabbala: Petrus Galatinus and Guillaume Postel," in *Political Hebraism: Judaic Sources in Early Modern Political Thought*, ed. Gordon Schochet, Fania Oz-Salzberger and Meirav Jones (Jerusalem and New York: Shalem Press, 2008): 3–28, here 16; Hartmut Bobzin, *Der Koran im Zeitalter der Reformation* (Beirut: Ergon i.K., 2008), 390, 399, 463.

22 Bernard Lewis, "On Occidentalism and Orientalism," in ibid., *From Babel to Dragomans: Interpreting the Middle East* (London: Weidenfeld & Nicolson, 2004): 430–38, here 432.

His criticism anticipated the debate on Orientalism that would begin about eighty years later.[23] Moreover, as Toner (2014) argues, the Arabs and the Orient functioned as a screen onto which ideas of freedom were projected.[24] Edward Gibbon (1734–1764?) considered the democracy of the Arabs as somehow being comparable to that of the ancient Greeks and the Romans, although it was based more on an individual love for freedom than on institutions. Processes of development in Arab society were accompanied by a minstrel poetry that fostered morality and modesty.[25] Cultural entanglements, as well as cultural delimitations, show that the modern Eastern Mediterranean has much in common with the West, while at the same time having in some respects a distinct nature. Contemporary academic approaches speak of "phantom borders" that still have an effect today and that stem from imperial boundaries long since vanished.[26]

The question of where Europe begins and ends has been the subject of different debates up until the present, while we must be aware that Europe as a continent and Europe as an idea are not identical. Any reflection on the nature of Europe will have to deal with its heritage, which goes well beyond its borders. We can therefore change our perspective and see Europe, as Nietzsche did, as a tiny island protruding from the vast continent of Asia.[27] Still, for historical reasons (see above), the term "Europe" is associated more with the Western part of the continent than with its Eastern part. Therefore, people tend to speak of "Europe" in Hungary and Russia,[28] and also in Greece, when in fact they mean France, Britain or some other country in *Western* Europe. The term "Levant," which, as I have shown elsewhere, originally denoted the entire Eastern Mediterranean, is similarly complex and ambivalent. Since it was closely related to the big port cities, "Levant" acquired the connotations of a mixed population (as is typical of port cities), a hybrid culture, and sectarian diversity. When a pro-

---

23 Kreutz, *Das Ende des levantinischen Zeitalters*, op. cit., 291. Gunnar Hering, "Die Auseinandersetzungen über die neugriechische Schriftsprache," in *Sprachen und Nationen im Balkanraum: Die historischen Bedingungen der Entstehung der heutigen Nationalsprachen*, ed. Christian Hannick (Cologne: Böhlau, 1987): 125–94, here 163 fn 164.; cf. Peter Mackridge, *Language and National Identity in Greece, 1766–1976* (Oxford: Oxford University Press, 2010): 275–7.

24 Toner, *Homer's Turk*, op. cit., 210, 212.

25 Osterhammel, *Die Entzauberung Asiens*, op. cit., 266–7.

26 See *Phantomgrenzen – Historische Grenzen als Thema der Grenz/Raumforschung*, ed. Sabine von Löwis (not yet published).

27 Friedrich Nietzsche, "Jenseits von Gut und Böse," in idem, *Sämtliche Werke: kritische Studienausgabe in 15 Bänden*, ed. by Giorgio Colli and Mazzino Montinari, Vol. 5: *Jenseits von Gut und Böse; Zur Genealogie der Moral* (Munich: DTV and De Gruyter, [1988/1999] 2012): 9–243, here 72.

28 Timothy Garton Ash, *Ein Jahrhundert wird abgewählt: Aus den Zentren Mitteleuropas 1980–1990* (Hamburg: Hanser, 1990), 163.

found transformation of the Eastern Mediterranean took place with the victory of the national idea, the newly founded nation states expelled many of their minorities, usually along sectarian lines, so that the term "Levant" was narrowed down to include what is today's Lebanon and Syria – two countries where the old Levant, which originally included Southeast Europe as well,[29] still seems to be alive today. Postcolonialism is unable to grasp such complexities and overlappings of identity since it remains within a binary view that seeks to praise the Islamic Middle East at the expense of Europe, with proponents of postcolonialism tending to play both entities off against each other. For the same reason, the discourse of enlightenment that took place in the Eastern Mediterranean from the early nineteenth century does not fit into postcolonialism's binary perception, since it was a shared phenomenon that transcended cultural and political borders. I will show in the next chapter how a different view may help us to a better understanding of the modern history of the Eastern Mediterranean.

1.2 Enlightenment in Perspective

Western ideas of enlightenment reached Southeast Europe as well as the Middle East at more or less the same time as a cultural revivalism that emerged in a later period of the Ottoman Empire.[30] This made the Ottoman Empire search for its own way of coping with the challenges posed by the threat of disintegration. According to Barqāwī (1988), the Ottomans were not simply anti-Western, but also anti-Eastern, therefore evading any East-West binarity.[31] This is important insofar as some scholars claim within the context of debates on the beginnings of modernity in the Islamic world that the Enlightenment was a worldwide phenomenon that took place in a similar matter and at roughly the same time, thereby denying any pioneering role to Western societies. This claim, and the debate on whether the Enlightenment occurred at all in non-Western societies and, if so, when it began, go back at least twenty years. Schulze (1990) has argued that the Islamic world not only underwent a process of enlightenment, but that this process had already begun in the eighteenth century, when the Enlighten-

29 Kreutz, *Das Ende des levantinischen Zeitalters*, op. cit., 17–28.

30 Matl, op. cit., 449.

31 إن الشرق في ظل العثمانيين هو غير الشرق الذي يجب أن يكون كما كان، إذ ذاك فليس العثمانيون نقيض الغرب فحسب، بل ونقيض الشرق معاً. ويتقلص الاختلاف بين الشرق العربي والغرب Aḥmad Nasīm Barqāwī, *muḥāwala fī qirā'at 'aṣr an-nahḍa: al-iṣlāḥ ad-dīnī, an-naz'a al-qawmīya* (Beirut: ar-Ruwwād li-n-nashr wa-t-tawzī', 1988), 45.

ment was in full swing in Europe. This argument has been strongly criticized particularly by Radtke (1994), and Hagen/ Seidensticker (1998).[32]

Although Schulze's critics were undoubtedly correct in pointing to all the flaws in his argumentation, they did not address the salient point that he was making: namely, that no historian had ever claimed that the Enlightenment had swept over the entire European continent and had affected all its countries with the same intensity and at the same time. On the contrary, it was often said that, for example, the Greek Enlightenment movement occurred roughly a century later than it had occurred in the Western part of the continent, and that, when it reached its peak in as late as the nineteenth century, it did so under the spell of French and British philosophers. The question to be asked should therefore be: why can any Enlightenment movement in the Islamic world not also be dependent on Western intellectual currents?

When Schulze accuses scholars of being affected by some kind of Eurocentrism, then this is therefore misleading, since no scholar of European history assumes that there ever took place at precisely the same time an overall, pan-European process of Enlightenment. In fact, the Enlightenment in Southeastern Europe evolved in a relationship of dependency on Western streams of thought – namely, from France, but also partly from the German-speaking areas. In other words: the Greek Enlightenment was no less dependent on Western Europe than the Middle Eastern and the Arab Enlightenment. Schulze's view of Europe as a continent of Enlightenment simply ignores the fact that much of Europe was – to put it pointedly – just as strongly influenced by French philosophers as the Middle East. Representatives of the Enlightenment both inside and outside Western

---

32 Reinhard Schulze, "Das islamische achtzehnte Jahrhundert. Versuch einer historiographischen Kritik," in *Die Welt des Islams*, vol. 30 (1990): 140 – 159, *passim*; idem, "Was ist die islamische Aufklärung?," in *Die Welt des Islams*, vol. 36 (1996): 276 – 325, here 230; idem, "Gibt es eine islamische Moderne?," in *Der Islam und der Westen: Anstiftung zum Dialog*, ed. Kai Hafez (Frankfurt/Main: Fischer 1997): 31 – 43, *passim*. Bernd Radtke, "Erleuchtung und Aufklärung: Islamische Mystik und europäischer Rationalismus," in *Die Welt des Islams*, vol. 34 (1994): 48 – 66, here 60; cf. idem, *Autochthone islamische Aufklärung im 18. Jahrhundert: theoretische und filologische Bemerkungen; Fortführung einer Debatte* (Utrecht: M.Th. Houtsma Stichting, 2000), *passim*; Gottfried Hagen and Tilman Seidensticker, "Reinhard Schulzes Hypothese einer islamischen Aufklärung. Kritik einer historiographischen Kritik," in *Zeitschrift der Deutschen Morgenländischen Gesellschaft*, Vol. 148 (1998): 38 – 110, *passim;* Jürgen Osterhammel, "Welten des Kolonialismus im Zeitalter der Aufklärung," in *Das Europa der Aufklärung und die aussereuropäische koloniale Welt*, ed. Hans-Jürgen Lüsebrink (Göttingen: Wallstein, 2006): 19 – 36, here 19 – 20; Christoph Herzog, "Aufklärung und Osmanisches Reich: Annäherung an ein historiographisches Problem," in *Die Aufklärung und ihre Weltwirkung*, ed. Wolfgang Hardtwig (Göttingen: Vandenhoeck Ruprecht, 2010): 291 – 321, here 294 – 7.

Europe were inspired more by French thinkers than by any other intellectual source.

This is all the more doubtful since, even within the Western European Enlightenment, the strength of influence between countries are not necessarily the same. Some scholars such as Helmuth Plessner have even argued that the Enlightenment never took root in German culture since it was rejected by Lutheranism, while France and England, with their "glorious revolutions," embraced the idea of a this-worldly rational state within a democratic-parliamentarian framework.[33] Thus, Schulze's claim that the entire European continent was affected evenly by the Enlightenment movement is flawed; also flawed is his conclusion that the Arab world must have participated in the Enlightenment. The only way to discover whether this was the case or not is to examine the historical facts.

Schulze's view (albeit, in a modified form) has prevailed, however, since it has become a common goal not only of Middle Eastern studies to account for the achievements of Western modernity within the context of the resources provided by Islam and Islamic cultures. Here another kind of criticism with a different starting-point comes into play, one that rejects all discussion of an Arab-Islamic world struggling with a modernization problem, and that considers such a discussion to be Eurocentric, irrespective of the most basic facts. In the words of Salvatore (1997):

> The major obstacle to seeing the making of an Arab-Islamic framework of reference as an original, though less spectacular token of modernity is precisely in the reconstruction of the Western model of Enlightenment modernity as a rather brusque passage, a clear break with "tradition." According to this heroized and simplified, but still solidly dominant view of the modern intellectual breakthrough, Arab-Islamic modernity could be at best assessed as being blocked at a proto-modern stage still waiting for an "enlightening" turn, at the threshold of a Lockian formulation of a community of free individuals as an actualization of Divine Will.[34]

This line of argumentation misses the point at more than one level. Its main problem is that it ignores the current state of affairs in the Arab-Islamic world, which should be analyzed and explained. The facts on the ground simply point in the opposite direction: that both the Eastern Mediterranean Orthodox world, as well as the Middle Eastern Muslim world, suffer from an absence of political stability, this absence being the result of internal developments.

---

33 Helmuth Plessner, *Die verspätete Nation* (Stuttgart et al.: Kohlhammer, 1959), 52, 56.
34 Armando Salvatore, *Islam and the Political Discourse of Modernity* (Reading: Ithaca Press, 1997), 50.

We must be aware that the modern state in the West is a major achievement that took decades to evolve into its principle form. As Fukuyama (2014) has pointed out, America lagged behind Western Europe in its creation of a modern state, since its political culture was from the very beginning skeptical towards the government.[35] America was a quasi-libertarian state up until the 1880s, when agrarian structures were predominant, but it improved its infrastructure under Woodrow Wilson, "the father of American public administration," when it decided to develop its economy.[36] But are apologists of Islam eager to maintain that the Islamic world has nothing to learn, nothing to reform – and nothing to overcome with the sole exception of the West's calamitous influence?

This is epitomized by the theory of Bauer (2011), who claims that the Enlightenment has never been an issue for the Islamic world since it has never had a need for it. According to Bauer, the Islamic world was for most of its existence characterized by its ability to endure and even cultivate ambiguity. This so-called "tolerance of ambiguity" means here that contradictions, counter-currents and discords within one and the same culture were not resolved to the benefit of one side or the other, but rather met with praise for being manifestations of a cultural and spiritual richness.[37] It therefore made any attempt at enlightenment null and void. Bauer connects his thesis with the assertion that the shortcomings of the contemporary Islamic world are phenomena that have mostly emerged within the last two hundred years or so, when Western advancement in the Middle East led to the adoption by local elites of Western values. These values were not only marred by a lack of ambiguity that ran counter to the traditional values of Middle Eastern Islamic societies; they also came into play at a time when the West was already a step further and was embracing new values. Bauer's argument is questionable in many respects and at more than one level, and it represents a new essentialism in Arabic and Islamic studies based as they are on bold interpretations and a decontextualized reading of sources.

This finds its logical end in modern-day attempts to craft theories of a global modernity (or even modernities), all of which boil down to the claim that every society in the world is in its entirety intrinsically modern and lives its own share of modernity. Thus, Kozlarek (2011) speaks of a "world consciousness" (*Weltbewusstsein*) as the center of modernity.[38] As I have shown, such attempts deliberately ignore any possible cultural differences, and selectively focus on the fact

35 Francis Fukuyama, *Political Order and Political Decay*, op. cit. 165.
36 Ibid., 150 – 2.
37 Thomas Bauer, *Die Kultur der Ambiguität*, op. cit., *passim*.
38 Oliver Kozlarek, *Moderne als Weltbewusstsein: Ideen für eine humanistische Sozialtheorie in der globalen Moderne* (Bielefeld: transcript, 2011), *passim*.

that contemporary societies are so entangled that they constitute something of a global market.[39] It is obviously not sufficient simply to diagnose the existence of a global modernity, since nations differ in their ability to create and develop modern technological goods. If we wish to discover when nations departed on different tracks, then we have to return to the past and look more closely at the nineteenth and early twentieth century.

Overcoming the metaphysical worldview is at the core of what is called in a Western European context "Enlightenment." Metaphysical thinking is deeply rooted in the idea that everything can be understood according to one principle. This idea was challenged when Descartes made the reasoning subject the center of perception.[40] The seventeenth century was a "metaphysical century" (Cassirer 1946) that the Enlightenment tried to break with. Ideas were no longer abstract, but became eligible for political discourse..[41]

The metaphor of light had been used in other contexts, too. For example, in the scientific literature of the classical age, beginning with sophism and Socrates, who declared the human being to be the measure of all things, before this approach gave way to a new mysticism.[42] We can also find in the Midrash Tanshuma, which was written in the Eastern Roman era, a variety of Greek loanwords such as פוטינוס (φωτεινὸς), which means "enlightened" and was used for people who had renounced idolatry. In Eastern Roman texts, the term φωτίζεσθαι (literally, "to become enlightened") means "to be baptized."[43] As Blumenberg has pointed out, from the antiquity, metaphysical thinking followed a cosmological pattern in which ideas constitute a cosmos as the model of the corporeal world. In this context, language and therefore rhetoric are a means less for communication than for achieving approval and tolerance.[44] The idea of a plurality of worlds, which started with William of Ockham, caused cracks to appear in the old metaphysical cosmology, and paved the way for an understanding of the world that saw it as being contingent and therefore open to human-driven progress. The Enlightenment took this as a starting-point to con-

---

39 Kreutz, *Zwischen Religion und Politik*, op. cit., 75–9.

40 Floris Cohen, *Die zweite Erschaffung der Welt* (Frankfurt/Main and New York: Campus, 2010), 154–5, 159.

41 Hans Blumenberg, *Arbeit am Mythos* (Frankfurt/Main: Suhrkamp, 1996), 25–6.

42 Wilhelm Nestle, *Griechische Weltanschauung in ihrer Bedeutung für die Gegenwart: Vorträge und Abhandlungen* (Stuttgart: Hannsmann, 1946), 54–5, 250.

43 Jakob Winter and August Wünsche, *Geschichte der jüdisch-hellenistischen und talmudischen Litteratur: Zugleich eine Antho*logie für Schule und Haus. Vol. 1, *Die jüdische Litteratur seit Abschluss des Kanons* (Trier: S. Mayer, 1894), 413–4.

44 Hans Blumenberg, *Wirklichkeiten in denen wir leben: Aufsätze und eine Rede* (Stuttgart: Reclam, 1999), 107–8.

front the human with the idea that the world that he or she had created was not the only one possible, and therefore to guide the human out of his or her "anthropocentric illusion" (Blumenberg) by confronting him with the possibility that the world he created need not be the only one.[45]

As with many other terms, the "Enlightenment" was coined in hindsight, with the eighteenth-century proponents of the French Enlightenment calling themselves *philosophes*. The notion of the Enlightenment seems to have been an invention of the nineteenth, one that replaced older notions and was promoted by Thomas Carlyle in his *History of the French Revolution* (1837). The English translator of Hegel's *Philosophy of History* had by the end of the nineteenth century used the French term *éclaircissement* on discovering that no appropriate term existed in English. The term "Scottish Enlightenment" did not appear until 1900. The *Encyclopaedia Britannica* used the term "Enlightenment" in 1929 without further specification and mainly for the German *Aufklärung*.[46]

In Scotland, the Enlightenment took place in a literary society where the general interest in progressive ideas was met with approval by a wider public.[47] As Herman (2001) has shown, there were in 1795 20,000 authors able to make a living from their writing in a Scottish population of 1.5 million.[48] Scotland was not likely to become a hotspot of progressive ideas when it stood in the shadow of its English neighbor to the South. As part of a twin kingdom after the downfall of the Roman Empire, Scotland had adopted the English language during the eleventh century at the expense of its native Gaelic.[49] The Scottish Enlightenment emerged in the context of what has been called "Reformed scholasticism,"[50] and was based very much on the idea of liberty being rooted in commerce and requiring a political frame of reference that owes to the individual.[51]Individual freedom and individual reasoning found their greatest

45 Hans Blumenberg, *Die Legitimität der Neuzeit* (Frankfurt/Main: Suhrkamp, 1996), 173.

46 Getrude Himmelfarb, *The Roads to Modernity: The British, French, and American Enlightenments* (New York: Vintage, 2004), 12.

47 Arthur Herman, *How the Scots Invented the Modern World: The True Story of How Western Europe's Poorest Nation Created Our World & Everything in it* (New York: Crown, 2001), 20.

48 Ibid., 21.

49 Ibid., 21, 23.

50 James Moore, "The two systems of Francis Hutcheson: On the origins of Scottish Enlightenment," in *Studies in the Philosophy of Scottish Enlightenment*, ed. by Michaael Stewart, Oxford: Clarendon Press, 2000: 37–60, here 39.

51 Fania Oz-Salzberger, "Freiheit und die ‚Gemeinschaft aller' in der schottischen Aufklärung," in *Kollektive Freiheitsvorstellungen im frühneuzeitlichen Europa (1400–1850)*, ed. Georg Schmidt, Martin van Gelderen, and Christopher Snigula (Frankfurt/Main et al.: Peter Lang, 2006): 419–29, here 420.

enemy in Scotland in fanaticism.[52] We can speak in a broader sense of a British Enlightenment in which Scottish philosophers were in the majority.[53] The British Enlightenment emphasized social virtues, and regarded reason as a means to an end rather than an end in itself.[54] Within this framework, it was the English approval of religion, commerce and freedom that found admiration among French philosophers such as Montesquieu.[55] In general, France was heavily influenced by English thinkers such as Bacon, Locke and Newton.[56] The term "Enlightenment" encompasses of course a diversity of opinions and approaches that had different implications. According to Himmelfarb (2004), the French Enlightenment was little more than a belated reformation, a search for reason. The emphasis on reason remains a specificity of the French Enlightenment. There is also diversity within the Enlightenment movement. According to Himmelfarb (2004), the British Enlightenment was based on virtue; the French, on reason; and the American, on liberty.[57] The Enlightenment had been the subject of public discourse in the American colonies from the beginning.[58]

Not unsurprisingly, the ideas of the Enlightenment were accompanied by an anti-colonial attitude in, for example, South America.[59] In Germany, it was August Ludwig Schlözer who championed the idea in his *Universal History* (1772–3) that the peoples of the world are not antagonistic but rather completing each other. One of his students, Johann Gottfried Herder, developed this idea further which had a significant impact on the emerging nationalist movements in the Eastern Mediterranean.[60] His thinking was the outcome of a universalist turn. Since it aimed to find the general laws of humankind and is universalist by definition, the era of the Enlightenment was universalist by nature.[61]

---

52 Herman, op. cit., 18.
53 Himmelfarb, op. cit., 5–6, 14.
54 Ibid., 19.
55 Ibid., 51.
56 Ibid., 5.
57 Ibid., 18–9.
58 Frank Kelleter, *Amerikanische Aufklärung: Sprachen der Rationalität im Zeitalter der Revolution* (Paderborn: Schöningh, 2002), 630.
59 Benedict Anderson, *Die Erfindung der Nation: Zur Karriere eines folgenreichen Konzepts* (Frankfurt/Main and New York: Campus, [2nd] 1993), 71.
60 Holm Sundhaussen, *Der Einfluss der Herderschen Ideen auf die Nationsbildung bei den Völkern der Habsburger Monarchie* (Munich: R. Oldenbourg, 1973), 64; Jörn Rüsen, *Kultur macht Sinn: Orientierung zwischen Gestern und Morgen* (Cologne et al.: Böhlau, 2006), 231.
61 Sibylle Tönnies, *Der westliche Universalismus: Eine Verteidigung klassischer Positionen* (Opladen: Westdeutscher Verlag, 1995), 21; Ernst Cassirer, *Nachgelassene Manuskripte und Texte*, ed. Klaus Christian Köhnke, John Michael Krois and Oswald Schwemmer, Vol. 3: *Geschichte. Mythos. Mit Beilagen: Biologie, Ethik, Form, Kategorienlehre, Kunst, Organologie, Sinn, Sprache, Zeit*, ed.

This calls for further explanation. Based on rationalism and knowledge, and fostering the idea of autonomous personalities, the Enlightenment was the unfolding of modernity. It sought to take away the fear from man and place him in a state of mastery, as Adorno and Horkheimer (2003) argued.[62] According to Max Weber's much-quoted phrase, the modern world is the product of disenchantment, beginning with the revolution of the Puritans, and followed later by those of the Enlightenment.[63] Disenchantment, as Adorno and Horkheimer understood it, is the overcoming of animism.[64] Blumenberg (2001) once claimed that the Enlightener was both afraid of the human and at the same time able to draw a lesson from humanity's errors, with these errors constituting a manual for those who seek knowledge.[65]

The Enlightenment had its downside, too, however. Isaiah Berlin (2002) called Rousseau "the most dangerous enemy of individual liberty, because he had declared that 'In giving myself to all, I give myself to none'."[66] Cassirer (1929), Koselleck (1959) and Bubner (1996) all agree that Rousseau sacrificed the individual for the sake of the absolutist will of the community, which he saw as being infallible.[67] As Sloterdijk (2007) reminds us, it was Rousseau who championed a strong link between the state and religion. However, when he failed in the face of middle-class and Catholic caveats, Rousseau adopted an increasingly anti-Christian stance. "Those who speak today of totalitarianism must not forget," says Sloterdijk, "that it had its dress rehearsal as revolutionary civil religion."[68]

Klaus Christian Köhnke, Herbert Kopp-Oberstebrink and Rüdiger Kramme (Hamburg: Felix Meiner, 2002), 77.

62 Theodor W. Adorno and Max Horkheimer, *Dialektik der Aufklärung* (Frankfurt/Main: Suhrkamp, [1947] 2003), 19.

63 Osvaldo Guariglia, *Universalismus und Neuaristotelismus in der zeitgenössischen Ethik* (Hildesheim et al.: Georg Olms, 1995), 104; Adorno and Horkheimer, op. cit., 19.

64 Ibid., 21.

65 Blumenberg, *Lebenszeit und Weltzeit*, op. cit., 199; cf. idem, *Höhlenausgänge* (Frankfurt/Main: Suhrkamp, 1996), 508–10.

66 Isaiah Berlin, *Liberty: Incorporating Four Essays on Liberty*, ed. Henry Hardy (Oxford and New York: Oxford University Press, 2002), 209–10.

67 Ernst Cassirer, *Die Idee der republikanischen Verfassung: Rede zur Verfassungsfeier am 11. August 1928* (Hamburg: De Gruyter, 1929), 11; Reinhart Koselleck, *Kritik und Krise: Ein Beitrag zur Pathogenese der bürgerlichen Welt* (Freiburg and Munich: K. Alber, 1959), 136–8; Rüdiger Bubner, *Welche Rationalität bekommt der Gesellschaft? Vier Kapitel aus dem Naturrecht* (Frankfurt/Main: Suhrkamp, 1996), 76.

68 Peter Sloterdijk, *Gottes Eifer: Vom Kampf der drei Monotheismen* (Frankfurt/Main and Leipzig: Verlag der Weltreligionen, 2007), 62.

The Enlightenment emphasizes reason, but reason is not what legitimates the Enlightenment, since, as Blumenberg pointed out, it cannot answer the question of why reason was inactive before Descartes, and why it came into being – at least not within the framework of its own notions.[69] Therefore, following Kant, we can understand the Enlightenment as an evolutionary process that aims to reform a person's mentality and as an individual and natural process that is necessarily more than simply toppling a tyranny.[70] Like Kant, we can define the Enlightenment as the attempt to fight for the freedom to use reason in the public domain.[71]

As such, the Enlightenment can be understood as a "perpetual council in which these articles are deliberated, fostered and defended against heretics" (Sloterdijk).[72] Although, as Flasch (1989) has shown, the medieval age was already familiar with criticism of other-worldly orientations, repressive sexual moralism and religious dogmatism,[73] matters came to a head during the Enlightenment. Berlin (1994) praised for good reason proponents of the Enlightenment such as Helvétius, Holbach and Condorcet, whom he praised as "great liberators" who "liberated people from horrors, obscurantism, fanaticism, monstrous views. They were against cruelty, they were against oppression, they fought the good fight against superstition and ignorance and against a great many things which ruined people's lives."[74]

Critics have tried to downplay the significance of modernity and therefore of the Enlightenment by seeing them as mere perpetuations of old theological views in a secular fashion. This criticism was strongly rejected by Blumenberg (1988), who emphasized the "legitimacy of the modern era."[75] We must note in any case that the Enlightenment in Western Europe was directed less against the medieval era and the church, and more than anything else against the absolutist state. According to Rhonheimer (2012), the absolutist state "cultivated an

---

69 Blumenberg, *Die Legitimität der Neuzeit*, op. cit., 159.
70 Immanuel Kant, "Beantwortung der Frage: Was ist Aufklärung" [1784], in idem, *Die Kritiken* (Frankfurt/Main: Zweitausendeins, 2008): 633–40, here 636, 639.
71 Ibid., 636.
72 Sloterdijk, op. cit., 187.
73 Kurt Flasch, *Aufklärung im Mittelalter? Die Verurteilung von 1277: Das Dokument des Bischofs von Paris übersetzt und erklärt* (Mainz: Dieterich, 1989), 13.
74 Isaiah Berlin and Ramin Jahanbegloo, *Den Ideen die Stimme zurückgeben: Eine intellektuelle Biographie in Gesprächen* (Frankfurt/Main: S. Fischer, 1994), 60–70.
75 Blumenberg, *Die Legitimität der Neuzeit*, op. cit., 107–8.

amalgamation of politics and religion which in this form was unknown to both the Christian antiquity and the medieval era."[76]

Here lies the key to understanding the rise of secularism in the West, since, as Fukuyama (1992) has argued, Christianity first had to secularize its objectives and in a sense to abolish itself before liberalism could be established.[77] It is important to keep this background in mind, since the prerequisites for any Enlightenment movement in the Eastern Mediterranean are so very different from those in the West at more than one level.

1.3 Theology and Science

It has long been claimed that the relation between state and religion is completely different in the case of Islam compared to the West, with the latter abiding by the "double order of state and church" (Rosenzweig).[78] A further difference, as Cook has put it, lies in the fact that "Christianity did not aspire to monopolize the domain of law in the way that Islam did."[79] For this reason, the law in Islam is in no way an ordinary matter.[80] Instead, the *sharia*, the law, is based on the conviction that the divine law marks the state of perfection and can therefore not be improved.[81]

The caliphate, which had been theoretically established between the ninth and the eleventh century, "embodied the Muslim intention to change the world in the name of truth. It kept the aura of a mission of salvation in Islam"

---

76 Martin Rhonheimer, *Christentum und säkularer Staat: Geschichte – Gegenwart – Zukunft* (Freiburg, Basel and Vienna: Herder, 2012), 351.

77 "Christianity in a certain sense had to abolish itself through a secularization of its goals before liberalism could emerge." Francis Fukuyama, *The End of History and the Last Man* (London: Hamish Hamilton, 1992), 216.

78 Franz Rosenzweig, *Der Stern der Erlösung* (Heidelberg: Schneider [3rd], 1954), chap. III,2, 117., cf. Salvatore: "The main difference between the Western and the Arab-Islamic historical paths is the less binding presence, in the second case, of a force-monopolizing entity demanding a special legitimation for itself (the modern state) in competition with a centralized instance of articulation of the Axial, cosmological and moral order (Church), as happened in Christian Europe." Salvatore, *Islam and the Political Discourse of Modernity*, op. cit., 53.

79 Michael Cook, *Ancient Religions, Modern Politics: The Islamic Case in Comparative Perspective* (Princeton/New Jersey: Princeton University Press, 2014), 300.

80 Ibid., 304.

81 Ibid., 277.

(Lapidus 1992).[82] A major shift occurred in the eleventh century, when many people in the Arab-Islamic world sensed that Islam was caught in a process of decline, which led to a deep crisis.[83] This crisis yielded a new approach to the essence of Islamic rule, with the focus shifting away from guiding the community under the banner of Islam to passing to every believer the responsibility for "commanding the good."[84] Islamic jurists found ways over time to adapt the *sharia* to every situation imaginable, but they never dared to introduce new principles of law. This was due to the lack of earthly authority. Only in terms of trade or peace agreements was coexistence based on man-made legal principles possible.[85] Ottoman objections to a human-made law were deep and emerged in as early as the sixteenth century.[86]

Western Europe underwent a process of individualization whose beginnings can be traced back to the fourteenth century, a process that encouraged people to think in less religious terms. We find this, for example, in Lorenzo Valla, who denied the existence of both divine and human freedom, but in other thinkers as well.[87] In contrast, the Orthodox world embarked on a different path and departed from the West in different stages. This began with the Council in Trullo in 692, which marked the beginnings of the schism.[88] It reconnected with a renewed Hellenism in 863, when Bardas re-founded the secular university of Constantinople.[89] And, more importantly, it was advanced by the teachings of Gregory Palamas, which became the official doctrine of the Orthodox Church in 1351. After Palamas was canonized in 1368, asceticism and prayer grew significantly in the Orthodox faith, when humanists in the West discovered the work of the

---

82 Ira M. Lapidus, "Islamisches Sektierertum und das Rekonstruktions. und Umgestaltungspotential der islamischen Kultur," in *Kulturen der Achsenzeit, Bd. II: Ihre institutionelle und kulturelle Dynamik, Teil 3: Buddhismus, Islam, Altägypten, westliche Kultur*, ed. Shmuel N. Eisenstadt (Frankfurt/Main: Suhrkamp, 1992): 161–88, here 168.

83 Tilman Nagel, *Die Festung des Glaubens: Triumph und Scheitern des* islamischen *Rationalismus im 11. Jahrhundert* (Munich: Beck, 1988), 203.

84 Ibid., 285.

85 Cook, *Ancient Religions*, op. cit., 278–9.

86 Ibid., 274–5.

87 Paul Richard Blum, *Philosophieren in der Renaissance* (Stuttgart: Kohlhammer, 2004), 52.

88 Christoph Dawson, *Die Gestaltung des Abendlandes: Eine Einführung in die Geschichte der abendländischen Einheit* (Leipzig: Jakob Hegner, 1935), 185.

89 Romilly J. H. Jenkins, *Byzantium: The Imperial Centuries, AD 610–1071* (Toronto: University of Toronto Press, 1987), 164; Dawson, op. cit., 178.

Greek Church fathers for themselves and incorporated the ancient Greek teachings into the Christian doctrine.[90]

The fixation on the other world is apparent in the writings of John of Damascus, who praised paradise on different occasions,[91] while downplaying the importance of life in this world,[92] which he saw as being imbued with impermanence.[93] For John of Damascus, the human was the master of the world,[94] and was expected to turn his face to God, while only sophistry regarded the human as the measure of all things.[95] The human will is limited and dependent on God's will.[96] In terms of liberty, John of Damascus argued that the will is the desire of someone who examines the things around him. Thus, the will is eager to examine things, or it is not. It can be subject to knowledge or not. Bringing the will into order transforms it into a free decision called choice. The decision is then the outcome of what is chosen and what is not chosen.[97]

As this may sound very simple, liberty and free choice have been less of an issue in Muslim history than reason and justice. In order to understand the challenges to Islamic theology and why so many thinkers believe in the need for religious reform, we should take a closer look at the situation beforehand. As Nagel (1988) has explained, Islamic theology is a complex issue with many facets, but light can nonetheless be shed on some major features. The Quran is based on the belief that God is the creator, that he is one and eternal. Islamic

90 Klaus Buchenau, *Auf russischen Spuren: Orthodoxe Antiwestler in Serbien, 1850–1945* (Wiesbaden: Harrassowitz, 2011), 24; Hans-Georg Beck, "Humanismus und Palamismus," in *Actes du XIIe congrès internatioonal d'études byzantines, Ochride, 10–16 septembre 1961*, tome 1, Beograd 1963: 63–82, *passim.* "

91 Mi. PG 94, 912, 5–913, 14; Mi. PG 96, 797, 28, after Franz Dölger, *Der griechische Barlaam-Roman: Ein Werk des Hl. Johannes von Damaskos* (Ettal: Buch-Kunstverlag, 1953).

92 Cf. John of Damascus on this world and the other world: Mi. PG 94, 821, 11; PG 94, 1564, 44; Mi. PG 94, 1357, 21, after Dölger, op. cit., 69.

93 John of Damascus on the impermanence of the earthy life: Greg. Naz. Mi. PG 35, 881, 25; Mi. PG 35, 881, 14, after Dölger, op. cit.

94 John of Damascus on the human as master of the world: Mi. PG 94, 921, 4; Mi. PG 94, 912, 2; Mi. PG 95, 148, 30; Mi. PG 96, 785, 38, after Dölger, op. cit.

95 Nestle, op. cit., 55.

96 John of Damascus on freedom of will: Mi. PG 94, 937, 28; Mi. PG 95, 148, 27, after Dölger, op. cit.

97 "... Βουλὴ δέ ἐστιν ὄρεξις ζητητικὴ περὶ τῶν ἐφ' ἡμῖν πρακτῶν γινομένη. Βουλεύεται γάρ, εἰ ὀφείλει μετελθεῖν τὸ πρᾶγμα ἢ οὔ. εἶτα κρίνει τὸ κρεῖττον καὶ λέγεται κρίσις. εἶτα διατίθεται καὶ ἀγαπᾷ τὸ ἐκ τῆς βουλῆς κριθὲν καὶ καλεῖται γνώμη. ἐὰν γὰρ κρίνῃ καὶ μὴ διατεθῇ πρὸς τὸ κριθὲν ἤγουν ἀγαπήσῃ αὐτό, οὐ λέγεται γνώμη. εἶτα μετὰ τὴν διάθεσιν γίνεται προαίρεσις ἤγουν ἐπιλογή. προαίρεσις γάρ [sic!] ἐστι δύο προκειμένων τὸ μὲν (recte: ἓν) αἱρεῖσθαι καὶ ἐκλέγεσθαι τοῦτο πρὸ τοῦ ἑτέτρου." Joh. Dam. Exp.: Mi. PG 94, 945, 10.

theology, as Nagel has pointed out, has attempted to shore up this belief against doubts of any kind.[98] The oldest sources on Muhammad date back to no earlier than 700 CE, so that the archiving of reports on his deeds and words did not take place until two or three generations after his death.[99]

According to ḤM Zaqzūq (1984), the discourse on modern Islam focuses on reason (*ʿaql*) which is said to be "one of the strongest agents of the Islamic religion." Therefore, the unity of religion and reason clearly works in favor of Islam.[100] Stemming from the *ijtihād*, the studies of the *uṣūl al-fiqh* are seen by Zaqzūq as a philosophical science (*ʿilm falsafī*) that began with Imām ash-Shāfiʿī and that had already emerged before Islamic thought came under the influence of Greek philosophy.[101] As Abū Zayd (1992) has shown, Shāfiʿī's view bolstered the Arabness of the Quran, which is the reason that he rejected the idea that there were words in the Quran of non-Arab origin. To overcome the factual correctness of this idea, he declared those words that appeared to be non-Arab to be Arab, too.[102]

For Shāfiʿī, the language of the Quran was bereft of foreign elements since it was based on the language of the Quraysh, who enjoyed hegemony over the Arabic language.[103] Behind the ideological rift concerning the use of language were two approaches: while Abū Ḥanīfa preferred the implicit attitude (*al-mawqif al-ḍimnī*), al-Shāfiʿī advocated the inseparability of expression and meaning (*al-talāzum bayna al-lafẓ wa-l-maʿnā*). Hence, the latter considered the Arabic language – or, more precisely, Qurayshite Arabic – to be an essential element in the structure of the text.[104] Thus, Shāfiʿī rejected Abū Ḥanīfa's proposal of reading the Fātiḥa within a prayer in Persian for those who were not able to recite the Quran in Arabic. This meant that any prayer in any other language other than Arabic would not be allowed, regardless of whether the person praying was

98 Nagel, *Die Festung des Glaubens*, op. cit., 120 – 1.
99 Ibid., 207.
100 Muḥammad Ḥamdī Zaqzūq, *dawr al-islām fī taṭawwur al-fikr al-falsafī* (Cairo: Maktabat Wahba, 1984), 18 – 24.
101 Ibid., 17 – 8.
102 والعلم به عند العرب كالعلم ... لسان العرب اوسع الالسنة مذهبا، واكثرها الفاظا، ولا يعلمه يحيط بجميع علمه انسان غير نبي، بالسنة عند اهل الفقه، لا تعلم رجلا جمع السنن فلم يذهب منها عليه شيئ al-Shāfiʿī, *ar-risāla*, Beirut s.a., 42, quoted after Naṣr Ḥāmid Abū Zayd, *al-Imām aš-Šāfiʿī wa-taʾsīs al-īdīyulūǧīya al- wasaṭīya* (Cairo: Sina li-n-nashr, 1992), 11.
103 Ibid., 15; cf. Tilman Nagel, *Geschichte der islamischen Theologie: Von Mohammed bis zur Gegenwart* (München: Beck, 1994), 163 – 4.
104 Abū Zayd, *al-Imām aš-Šāfiʿī*, op. cit., 20.

able to read Arabic or not. When Shāfiʿī defined the preconditions under which a prayer was valid, he made clear that the prayer had to be in Arabic.[105]

Shāfiʿī made the analogy (*qiyās*) the only tool for discovering in the scripture the implicit (*mustatir*) meaning (*dalāla*), with meaning falling into two categories: *dalālat ibāna* and *dalālat ishāra.*[106] Therefore, as Abū Zayd goes on to argue, Shāfiʿī's extension of the competence of the *qiyās* is nothing but an ideological reinvention of textual authority, and that is the reason that he championed the *qiyās* over the *istiḥsān* (legal opinion). In Shāfiʿī's view, the *qiyās* is based on sound principles, while the *istiḥsān* is not; the *qiyās* seems to be a valuable protector (*ʿāṣim*) against incongruity (*khilāf*), which is the reason why he polemicized against intellectual and juridical pluralism (*at-taʿaddudīya al-fikrīya wa-l-fiqhīya*) of his time..[107]

Muḥammad b. Idrīs ash-Shāfiʿī (767–820), a leading representative of the age of codification (*tadwīn*),[108] contributed to the forming of Sunni Islam based on scholarly consensus (*ijmāʿ*).[109] Shāfiʿī's notion of *ijmāʿ* is in turn tied to the Prophet's companions and their agreement or disagreement in matters of faith.[110] The *ḥadīth* becomes the ultimate tool in this process, so that obedience to the prophet meant ultimately obeying the *sunna* (*ḥadīth*). He therefore made the *sunna* a divine revelation that had the same legislative power and compulsion as the Quran, that could not be defeated. Moreover, there was no longer a clear difference between the *sunna* of revelation and the *sunna* of habits and traditions. Shāfiʿī expanded the term *sunna* to include sayings, doings and authorizations (*muwāfaqāt*), which he sought to protect from criticism by binding it to the idea of *ʿiṣma* (infallibility), itself a feature of all prophets, and especially of Muḥammad. Meanwhile, the human nature of the prophet disappeared.[111]

As Abū Zayd goes on to argue, legitimizing the *sunna* through the claim that it was built on the interpretation (*ta'wīl*) of some texts in the Quran was not an act performed in isolation which Shāfiʿī misunderstood when he stuck with the *sunna* where it is not in accordance with the Quran as well as his denial that the *sunna* was a revelation (*waḥy*) from God.[112] Shāfiʿī's attempt to blend the revela-

---

105 Ibid., 18–9.
106 Ibid., 95.
107 Ibid., 101–2, cf. ibis. 110.
108 Ibid., 58.
109 Ibid., 93.
110 Ibid., 75–85.
111 هكذا يكاد الشافعي يتجاهل "بشرية" الرسول تجاهلا شبه تام، وتكاد تختفي من نسقه الفكري "انتم اعلم بشئون ديانكم"، حتى أنه يجعل من مواضعات النظام الاجتماعي السائد، والذي لم يقُمه الإسلام، سُنة واجبة الإتباع، يجري عليها القياس Ibid., 40.
112 Ibid., 42.

tion of the Quran (*waḥy al-Qur'ān*) with the revelation of the *sunna* (*waḥy as-sunna*) is not built on firm ground, since it comes close to blurring the distinction between the divine (*ilāhī*) and the human (*basharī*), the consequence of which was that, in depicting the prophet as someone who transmitted and explained the revelation, ignored both his exclusiveness (*khuṣūṣīya*) and his humanness (*basharīya*).[113]

For Shāfiʿī, the Quran and the *sunna* constituted one single text, which means that either text can abrogate the other. Yet, this is not what Shāfiʿī intended, since he abrogation only under the precondition that each of the two texts is independent.[114] Also, the Quran cannot abrogate the *sunna*; only the *sunna* can abrogate the *sunna*. But the *sunna* can clarify which elements in the Quran abrogate, and which can be abrogated.[115]

As Abū Zayd goes on to explain, it was during the time of Shāfiʿī that the *sunna* became the second source of legislation. However, it would be too simple to describe this process as a mere act of resistance to the *ahl al-ra'y* because, with regard to the legitimacy of the *sunna*, there was no real difference between the two and the *ahl al-ḥadīth*. Rather, they differed in terms of the trustworthiness of some kinds of *ḥadiths*. In summary, the *sunna* relates to the Quran in three respects: first, in its affirmative similarity (*tashābuh dalālī*) i.e., the *sunna* retells what is said in the Quran; second, in explaining and elucidating the Quran (*at-tafsīr wa-l-bayān*); and, third, in helping create a legislative text on its own.[116]

According to Abū Zayd, it was the third feature that became the bone of contention, i.e., the *sunna*'s ability to form legislation independently (*istiqlāl al-sunna bi-t-tashrīʿ*), which "has put on it the dust of forgetting our culture and our religious thought" (*alladhī ahyala ʿalayhi turāb an-nisyān fī thaqāfatinā wa-fikrinā d-dīnī*). This is why the *sunna* and the revelation became the source not of legislation, but of explication and elucidation (*at-tafsīr wa-l-bayān*) of what the Quran (*al-kitāb*) had only written in summary form. Abū Zayd argues that this stood in stark contrast to Shāfiʿī's own point of view, which was that the *sunna* was a kind of revelation (*waḥy*) that differed from the revelation of the Quran. Instead, the *sunna* was the "diction of fright" (*ilqā' al-rawʿ*), i.e., "revelation" in its lexicographical (*lughawī*), and not in its conventional (*iṣṭilāḥī*), meaning.[117]

---

113 Ibid., 46.
114 Ibid., 46–7.
115 Ibid., 68.
116 Ibid., 37–8.
117 Ibid., 39.

Abū Zayd concludes that Shāfiʿī took a middle position between the *ahl al-raʾy* and the *ahl al-ḥadīth*, since there was no real difference between them regarding the legitimacy (*mashrūʿīya*) of the *sunna*. Shāfiʿī had originally belonged to the *ahl al-ḥadīth*, but he recognized the legitimacy of the analogy (*qiyās*), which he then sought to shackle in a number of ways up to the point when they were only indirectly based on the texts. He then extended the scope of the texts to include both the *sunna* and the *ijmāʿ*, and only had to ensure that the *sunna* was regarded as a revelation.[118] As Abū Zayd stresses, the question of the rank of the *ḥadīths* and the importance of the *raʾy* were mere symptoms of a more profound struggle between the forces of change and progress, and those of inertia and hegemony.[119]

Abū Ḥanīfa saw the *sunna* as a secondary source that was meant only to elucidate and interpret the original text, the Quran.[120] In contrast to Abū Ḥanīfa, who followed the principle of *istiḥsān* (arbitrary personal opinion), and al-Shāfiʿīs' teacher Mālik b. Anas, who followed the principle of *al-maṣāliḥ al-mursala* (the interests left open by God), Shāfiʿī made the "Arabness" (*ʿurūba*) of the Quranic text his guiding principle.[121] He therefore accepted the *marāsīl*, i.e., uncertain oral traditions that presumably dated back to the Prophet, since, as Abū Zayd claims, the objective of his approach was "to form the memory on the basis of conservation."[122] This went together with the *ahl al-ḥadīth*, which defended the sovereignty (*ḥākimīya*) of the texts and their holism in terms of every sphere of human activity, while the *ahl ar-raʾy* defended reason (*ʿaql*) as represented in understanding the *al-maṣāliḥ al-mursala*, the *istiḥsān*, and the *istiṣlāḥ*. It was ultimately a struggle between *naql* and *ʿaql*.[123]

Abū Zayd goes on to argue that *sunna* meant for Shāfiʿī the ways of *aʿrāf* (customary laws), *ʿādāt* (customs), and *taqālīd* (traditions), and is not restricted to the narratives of the Prophet. That said, we understand why, with the extension of the notion of *sunna*, collecting *ḥadīths* grew in importance so as to

---

118 Ibid., 50; about his middle position between the *ahl ar-raʾy* and the *ahl al-ḥadīth* see also ibid., 56–7.

119 Ibid., 57.

120 Ibid., 56.

121 لأنه لا يعلم من إيضاح جمل علم الكتاب أحد جهل سعة لسان العرب وكثرة وجوهه، وجماع معانيه وتفرقها Shāfiʿī, *ar-risāla*, 50, quoted after Abū Zayd, *al-Imām ash-Shāfiʿī*, op. cit., 22–3.

122 Ibid., 74–5. Cf. entry *mursal* in EI (2), vol. 7, 631. "At first there was still difference of opinion on whether or not certain *mursalāt* or marāsīl traditions of certain major Successors constituted an argument (*ḥudjdja*) in a legal argument. It was finally al-Shāfiʿī who settled the matter: they do not, unless substan tiated by a Prophetic tradition of more or less the same purport supported by an uninterrupted (*muttaṣil*) and sound isnād."

123 Abū Zayd, *al-Imām ash-Shāfiʿī*, op. cit., 58.

leave less room for *ijtihād*, which was the primary goal (*al-ghāya al-asāsīya*) of Shāfiʿī's theology.[124] According to Abū Zayd, Shāfiʿī founded a heuristic principle in which the text provides a solution to every problem on a rationalist basis and at the same time at the expense of reason: the abolition of reason (*ilghāʾ al-ʿaql*) was achieved through reason.[125] In other words, Shāfiʿī tried to make use of rational tools in order to justify the nullification of reason.[126] This seems to be somewhat disturbing, since, according to the Quran, the human is blessed with the ability to think (6:104, 16:78) with "wisdom" (*ḥikma*) (2:269).[127]

Consequently, for the Shāfiʿī school, faith as prescribed by the Quran is a "burden" that the believer has to accept,[128] and the *sharia* chiefly means the "subjugation" of believers.[129] Thus, every detail of everyday life had to comply with the divine law.[130] According to the Shāfiʿī school, the *sharia*, usually defined as the body of normative texts that draw on God's will and the polity of believers in all its aspects,[131] is beyond any substantiation by human reason.[132] This is in contrast to the older school of Abū Ḥanīfa, which viewed the freedom of the human as something original, and slavedom (*ʿubūdīya*) as something temporary.[133] It was for this reason that Abū Ḥanīfa rejected some *ḥadīths* that ran counter to his ideas, e. g., the *ḥadīth* describing when the Prophet encountered six slaves and, because he did not have enough money to buy all of them, he freed two and sent the other four back into slavery. Abū Ḥanīfa rejected this because the foundations of religion (*uṣūl*) are conclusive, and any isolated tradition is only hypothetical; thus, for him, liberty (*ʿitq*) was the general solution for these slaves.[134]

In contrast, ash-Shāfiʿī believed in the institution of slavery since a *ḥadīth* commands that *man bāʿa ʿabdan wa-lahu māl fa-māluhu li-l-bāʾiʿ illā an yashtariṭahū al-mubtāʿ*, i.e., "if a person buys a slave who owns property, then the property belongs to the slave's owner as long as the buyer does not make it a prereq-

---

124 هو مفهوم الذي يتسع لدى الشافعي لسنن الأعراف والعادات والتقاليد، ولا يقتصر ـ كما سلفت الإشارة ـ على المروى عن الرسول وحيا وتشريعا ... وإذا يتسع نطاق السنة، يتسع مجال النصوص، وتضيق لذلك مساحة الاجتهاد Ibid., 89.

125 Ibid., 22.

126 حاول استخدام بعض آليات التفكير العقلي ليبرر نفي العقل Ibid., 90.

127 Cook, *Ancient Religions*, op. cit. 173; Muḥammad Sharafī [= Mohamed Charfi], *al-Islām wa-l-ḥurrīya: sūʾ at-tafāhum at-tārīkhī* (Tunis: Dar Petra, 2008), 127.

128 Nagel, *Die Festung des Glaubens*, op. cit., 224.

129 Ibid., 237.

130 Ibid., 204.

131 Ibid.

132 Ibid., 339.

133 Abū Zayd, *al-Imām aš-Sāfiʿī*, op. cit., 42.

134 Ibid., 61.

uisite." For Abū Zayd, introducing this saying into the *sunna al-waḥy* strongly opposes the goals of the *sharia* (*al-maqāṣid al-kullīya li-sh-sharīʿa*), which regards freedom (*ḥurrīya*) as original (*aṣlan*), and slavedom as temporary (*ṭāriʾan*).[135] Abū Zayd concludes that Shāfiʿī's ideology corroborates the rejection of a free human will.[136]

As GE von Grunebaum (1956) has pointed out, values such as the emphasis on otherworldliness and the priority of the community over the individual remained unquestioned and unchallenged even after the influx of Greek science: "Hellenic thinking in no way affected the vantage point or objective of Islamic thinking."[137] This is partly due to the Ashʿarī school, which merged with the Shāfiʿī school,[138] and created a theology based on God and his attributes without further modalities (*bilā kayfa*) – all translated into the rational language of Aristotelian logic. Hence, everything in the world was a mere expression of God's will, which left no space for freedom of thought.[139] For the Ashʿarite theologian al-Juwaynī, the human was free but placed under the burden of faith, with all his actions being predetermined by God.[140] This reflects the Quranic tendency of, as Lapidus (1992) has put it, "taking the world as it is and modifying it, rather than radically challenging and changing the world."[141]

In contrast to the Ashʿarīte theological school, Muʿtazilite metaphysics believed in personal responsibility for earthly matters.[142] According to the Muʿtazilite school, believers would be more flexible in the *sharia*,[143] since they believed that God only wanted the best for his creation. This was dismissed by critics as being an attempt to subordinate God to human interests.[144] The Ḥanīfa school

135 Ibid., 40 – 1.

136 Ibid., 24.

137 Gustave E. von Grunebaum, "The Problem of Cultural Influence," *in Charisteria Orientalia praecipue as Persiam pertinentia*, ed. Felix Tauer, Věra Kubíčková and Ivan Hrbek (Prague: Nakl: Ceskoslovenské akademie věd, 1956): 86 – 99, here 88 – 9, 91, 95.

138 Gerhard Endreß, "Der Islam und die Einheit des mediterranean Kulturraums im Mittelalter," in *Das Mittelmeer: Die Wiege der europäischen Kultur*, ed. Klaus Rosen (Bonn: Bouvier, 1998), 270 – 95, here 83.

139 Hava Lazarus-Yafeh, "Die islamische Reaktion auf den Rationalismus," in *Kulturen der Achsenzeit, Bd. II: Ihre institutionelle und kulturelle Dynamik, Teil 3: Buddhismus, Islam, Altägypten, westliche Kultur*, ed. Shmuel N. Eisenstadt (Frankfurt/Main: Suhrkamp, 1992): 210 – 25, here 216 – 7.

140 Nagel, *Die Festung des Glaubens*, op. cit., 228.

141 Lapidus, "Islamisches Sektierertum," op. cit., 165; cf. Quran 3,133 ("...sāriʿū ilā jannatin ...").

142 Nagel, *Die Festung des Glaubens*, op. cit., 117.

143 Ibid., 232.

144 Ibid. 226.

and the Mu'tazila formed for some time a theological coalition, but this coalition almost disappeared in the tenth century.[145] The major achievement of the Mu'tazila was the centrality of reason as a "source of religious knowledge, a touchstone of religious truth" (Goldziher 1970), which is the reason that modern reformists often draw on to their line of argumentation.[146]

Similarly, the philosopher Abū al-'Alā' al-Ma'arrī (973–1058) argued for the central importance of using reason, since "reason is a prophet." This use is driven by the force of doubt directed at all aspects of history and society.[147] When Ma'arrī declared that he was no infidel and that all he wished for from God was his own death (*mā anā bi-l-mulḥid al-kafūr wa-lā aṭlubu min rabbī ghayra ilḥādī*), and therefore refuting the claims that he had left Islam, he was playing on the different meanings of the word *ilḥād*, which can denote both "atheism" and "death," and which echoes the term *mulḥid* ("atheist"). It was not without a sense of irony that he boasted that the grave (*malḥad*) would be his last home, again playing on the word stem *l-ḥ-d*.[148] On the other hand, since God is omnipotent in al-Ma'arrī's theology, there is no free will left for the human, whose soul is the source of all evil.[149]

The only philosopher who outlined an individual soul as being independent of the "active intellect" was Abū l-Barakāt al-Baghdādī (d. 1164). His views are in stark contrast to, for example, al-Fārābī's system of the emanation of intellects. The source of human cognizance is in al-Baghdādī's philosophy no longer located in the divine *amr* (will), which is mentioned in the Quran, but in the autonomous being void of any ties to transcendency. However, al-Baghdādī's philosophy stood alone and had little to no impact on other philosophers. It was for this reason that he was given the byname "awḥad az-zamān" (the only of his time).[150] Al-Baghdādī would become a major influence on the Egyptian Enlightenment thinker Ṭāhā Ḥusayn and other rationalists in as late as the twentieth century.[151]

For Barqāwī (1988), religious reform or the attempt at modernizing Islam (*'aṣranat al-islām*) first occurred in the nineteenth century and during the age

145 Endreß, "Der Islam und die Einheit ...," op. cit., 83.

146 Ignaz Goldziher, *Die Richtungen der islamischen Koranauslegung* (Leiden: Brill, 1970), 136.

147 Muḥammad Abū al-Fadl Badran, "'...denn die Vernunft ist ein Prophet' – Zweifel bei Abū 'l-'Alā' al-Ma'arrī," in *Atheismus im Mittelalter und in der Renaissance*, ed. Friedrich Niewöhner and Olaf Pluta (Wiesbaden: Harrassowitz, 1999): 61–84, here 67.

148 Ibid., 68. The latter is from his *al-Fuṣūl wa-l-ghāyāt.*

149 Pierre Cachia, *Ṭāḥā Ḥusayn: His Place in the Egyptian Literary Renaissance* (London: Luzac, 1956), 170.

150 Nagel, *Geschichte der islamischen Theologie*, op. cit., 194–5.

151 Lazarus-Yafeh, "Die islamische Reaktion ...," op. cit., 218.

of the Nahḍa, when intellectuals became aware of the superiority of the West in both technical and political matters. These intellectuals responded by creating a "new Islamic consciousness" (*waʿy islāmī jadīd*),[152] one of whose trailblazers was the Egyptian theologian ʿAlī ʿAbdarrāziq.[153] There is a widespread conviction that so much in the Arab world is either a reaction to the West or the result of Western influence, which led one observer of Egyptian matters to conclude that, behind Muḥammad ʿAbduh, there is Martin Luther.[154] This contradicts Barqāwī, who, as a Marxist, holds responsible for the "bourgeoisification" of Islam (*barjazat al-islām*) internal factors of society.[155]

As Barqāwī notes, the central question for all these reformers was how to overcome the opposition between science and Islam. Hence, they had to find science and its importance in the Quranic text and the *hadith*. They viewed science (knowledge) as one of the most powerful ways to achieve progress, and there are in fact enough Quranic verses to corroborate this view. However, for Barqāwī, the reformers connected progress and science in a way that was too mechanical, and thereby located the core of Western progress within science. Since the West does not have a monopoly on science, they tried to make Islam the starting-point for their own scientific progress.[156]

1.4 Agents of Reform

While on the one hand Western scholarship emphasizes the importance of cultural entanglement, it is often assumed that the Western influence on Arab philosophical thought is more negative than positive, since it imposes patterns of Orientalism and colonialist preconceptions. This is somewhat surprising since, in the case of European countries such as Germany and Greece, this influence is not seen as being equally strong. As a matter of fact, mutual influence and perception have a long history that can be traced back to the early age of Islam before the Islamic world opened up to Western thought and literature in the nineteenth century – as the Christian societies in the Balkans also did.

The Eastern Church has a long tradition of dealing with Islam. Early scholars in the ninth century such as John of Damascus and his student Abū Qurra, who

152 Barqāwī, *muḥāwala*, op. cit., 61–2.
153 Ibid., 67.
154 Ibid., 64–5.
155 إن العروي ... لا يرى الأسباب الداخلية العميقة والجوهرية التي دفعت بواحد كمحمد عبده أو العظم أو رشيد رضا إلى برجزة الإسلام Ibid., 64.
156 Barqāwī, *muḥāwala*, op. cit., 51, 52.

wrote polemics against Islam, while also respecting it for being not merely a distorted version of Christianity but a religion of its own. The Eastern Roman monk Euthymios Zygabenos took a different approach when, in his theological treatise the *Panoplia*, he treated Islam as a heresy. He also gave a first account of the institution of the *jihad*.[157] The emperors Manuel II and others accepted that Islam was more than a superstition, and that it was instead an intellectual challenge to be tackled. This was part of a wider phenomenon since the controversies during this age were immense. Beginning with theology, these controversies soon spilled over into topics in the field of philosophy.[158]

In 1354, when the Greek theologian Gregory Palamas was in Ottoman captivity, he learned about Islam and became engaged in religious disputes.[159] Later on, it was the Ottoman threat that formed the background for the Reformation and that helped it to emerge.[160] The last of the great Eastern Roman philosophers, Georgios Gemistos Plethon (1355 – 1452), was impressed by the ethical conceptions of Islam, which, as a neo-Platonist, he turned against Christianity. As an admirer of ancient pagan philosophy, he attempted to invent a completely new religion, one that would have opposed to a certain degree the dogmas of Christianity.[161]

Although the relationship between Eastern Roman learning and non-Christian systems of thought was not an easy one, intellectual stimulation from other sources remained a vibrant part of Greek and Eastern Roman history up until the Ottoman period, when a new influence came into play. Having preserved the national cultures for centuries, the Orthodox Churches had to cope with a new challenge, one represented partly by Orthodox clerics who wanted to renew the Church and who therefore translated mainly French and German scriptures of the Enlightenment into Greek. The fact that the first generation of

---

157 Klein, "Hugo Grotius' Position on Islam," op. cit., 150 – 1; cf. Julia Gauß, "Glaubensdiskussionen zwischen Ostkirche und Islam im 8. – 11. Jahrhundert," in *Theologische Zeitschrift*, vol. 19 (1963): 14 – 28, *passim*, and Berthold Altaner, "Zur Geschichte der anti-islamischen Polemik während des 13. und 14. Jahrhunderts," in *Historisches Jahrbuch*, vol. 56 (1936): 229 – 30, *passim*.
158 Hans-Georg Beck, *Das byzantinische Jahrtausend* (Munich: Beck, 1994), 309.
159 M. Balivet, "Culture ouverte et échanges inter-religieux dans les villes ottomanes du xive siècle," in *The Ottoman Emirate* (1300 – 1389), ed. E. Zachariadou (Rethymnon: Crete University Press, 1993), 2 – 3, quoted after Colin Heywood, "The Frontier in Ottoman History: Old Ideas and New Myths" in idem, *Writing Ottoman History: Documents and Interpretations* (Aldershot: Ashgate, 2002), I 238 – 9.
160 Thomas Kaufmann, *Der Anfang der Reformation: Studien zur Kontextualität der Theologie, Publizistik und Inszenierung Luthers und der reformatorischen Bewegung* (Tübingen: Mohr Siebeck, 2012), 102 – 3.
161 Beck, *Das byzantinische Jahrtausend*, op. cit., 310.

translators and advocates of Western Enlightenment were clerics explains why the modern Greek Enlightenment and the Enlightenment in Southeast Europe in general never adopted an attitude as fiercely anti-clerical or anti-religious as in France.[162]

At the beginning of the seventeenth century, when Western powers such as France, Austria, England and the Netherlands were increasing their presence in the Eastern Mediterranean, both the French Catholic envoy, as well as his Dutch Calvinist counterpart, tried to convince the Ecumenical Patriarchate of Constantinople to adopt their respective theology. Influenced by the Dutch envoy Cornelis Haga (1578–1654), the Greek Patriarch Kyrillos Loukaris ordered Maximos of Gallipoli to translate the New Testament into contemporary (so-called Demotic) Greek, before the translation was printed in Leiden (the Netherlands) in 1683.[163]

A cultural revival among the Greek inhabitants of Southeast Europe began in as soon as the seventeenth century, when the neo-Aristotelian scholar Theofilos Koridalleas (1570–1646) became head of the Patriarchal Academy of Constantinople under Patriarch Kyrillos Loukaris in 1622, and the first printing press (reportedly brought into the country by Mitrofanis Kritopoulos) was established five years later. These events marked a new era of scholarship in many fields, including mathematics and medicine, but also theology and philosophy.[164]

The Phanariotes, a circle of intellectuals associated with the Patriarchate and committed to the idea of "illuminating the nation" (τὸν φωτισμὸν τοῦ γένους), began their reign in Moldavia and Wallachia between 1711 and 1716. In cooperation with the Patriarchate, they tried to Hellenize the local population.[165] Commerce had contributed to the Greek connection with Europe since the middle of the seventeenth century, and it became a monopoly of sorts for the Greeks within the Ottoman Empire. The newly accumulated wealth facilitated the de-

---

162 Matl, op. cit., 449–50.

163 Friedrich Heyer, *Die Orientalische Frage im kirchlichen Lebenskreis: Das Einwirken der Kirchen des Auslands auf die Emanzipation der orthodoxen Nationen Südosteuropas 1804–1912* (Wiesbaden: Harrassowitz, 1991), 2.

164 Gunnar Hering, *Ökumenisches Patriarchat und europäische Politik, 1620–1638* (Wiesbaden: Steiner, 1968), 159; Konstantinos Diovouniotis, *Μητροφάνους Κριτοπούλου ανέκδοτος γραμματική της απλής ελληνικής* (Athens: Makris, 1924), 121–2; Heyer, *Die Orientalische Frage*, op. cit., 2.

165 Hassiotis, I. K., "From the ‚Refledging' to the ‚Illumination of the Nation': Aspects of Political Ideology in the Greek Church under Ottoman Domination," in *Balkan Studies*, Vol. 40, No. 1 (1999): 41–55, here 52; Kemal Karpat, "Millets and Nationality: The Roots of the Incongruity of Nation an State in the Post-Ottoman Era," in *Christians and Jews in the Ottoman Empire*, ed. Benjamin Braude and Bernard Lewis (New York: Holmes and Meier, 1982), reprinted in Kemal Karpat, *Studies on Ottoman Social and Political History: Selected Articles and Essays* (Leiden et al.: Brill, 2002): 611–46, here 635.

mand for education.[166] Commercial ties with the German-speaking countries contributed to the rise of a Greek, and mainly Macedonian, middle class, which fostered education and the idea of liberty.[167]

The 1718 Treaty of Passarowitz between Habsburg and the Ottoman Empire ending the Austrian-Ottoman and Venetian-Ottoman wars allowed free shipping to each side of the Danube. When the Austrian Oriental Company was founded in 1719, commercial ties between the two countries were strengthened, and the Greeks in Macedonia were granted the right to settle in Austria and to become naturalized. There was as a consequence a steady flow of Greek migrants to Vienna, but also to other German-speaking cities such as Breslau and Chemnitz. By the end of the eighteenth century, there were more than 80,000 Greeks living in Austria alone. It was not only the Greeks that Austria had strong commercial ties with; rather, Austria maintained a whole network consisting of Christian minorities throughout the Ottoman Empire. It was at that time Trieste that had become the new hub of commerce and consequently a rival of Venice.[168]

Vienna was a magnet for diaspora Greeks and maintained strong ties with the Greek community of Leipzig.[169] We find in Germany eminent Greek scholars such as Eustratios Argentis (ca. 1685/90 – 1756/60), who possibly studied in Halle in about 1720 before returning to the island of Chios.[170] The Archimandrite Theokletos Polyeidis began his journey through Germany in 1727, where he travelled to Hesse and Wurttemberg before founded a Greek parish in the city of Leipzig in 1743. He then continued his travels to Berlin and finally arrived in Halle in about 1746. Leaving Germany for Russia, he translated into Greek a treatise purportedly by an Eastern Roman monk called Agathangelos, which appeared in 1751. In fact, the author was Polyeidis himself, and he drew in his treatise on the history of the Eastern Roman Empire while propagating an ardent nationalism with an admix-

---

166 Von Maurer, *Das griechische Volk*, Erster Band [1st vol.], op. cit., 19 – 20; Jonathan I. Israel, *Enlightenment Contested: Philosophy, Modernity, and the Emancipation of Man: 1670 – 1752* (Oxford: Oxford University Press, 2006), 318.

167 Apostolos Vakalopoulos, *Griechische Geschichte von 1204 bis heute* (Cologne: Romiosini, 1985), 56.

168 Ibid., 55. Rudolf Lill, *Geschichte Italiens vom 16. Jahrhundert bis zu den Anfängen des Faschismus* (Darmstadt: WBG, 1980), 33

169 Frank-Thomas Suppé, "In Sachsen auf Heimatboden. Zur Geschichte der griechischen Gemeinde in Leipzig von ihren Anfängen bis nach 1945," in *Eugénios Búlgaris und die griechische Aufklärung in Leipzig: Die Griechen im Leipzig des 18. Jahrhunderts; Eine Ausstellung der Universitätsbibliothek Leipzig vom 16. Bis 30. Oktober 1996* (Leipzig: Universitätsbibliothek, 1996): 13 – 48 here 24, 35.

170 Podskalsky, *Griechische Theologie*, op. cit., 331.

ture of pro-Russian and anti-Roman slogans. Although written in a more ancient Greek, the treatise is still popular in Greece today.[171]

A feature of this epoch is that we find reformist ideas on the one hand and nationalism on the other – side by side, and sometimes overlapping. In general, it is remarkable that the clerics took part in this discourse, going along sometimes with progressive ideas, and sometimes with more conservative ideas. This was against the background of strong anti-Western traits within the Orthodox Church up until then. This was linked to the mystical tradition within the Orthodox Church, often referred to as "Palamism," which dated back to Gregory Palamas. His school of thought was first brought into the Church with the synod of 1351 and the canonization of Palamas in 1368. According to Podskalsky (1988), the teachings of Palamas were an impediment to the ideas of the Enlightenment.[172] Yet, the term "Palamism" is disputed. As Nikolaou (1990) argues, Orthodox theology had neglected its own roots and become drawn into the disputes of reformation and counter-reformation under Ottoman imperialism, before finding its way back to its origins, and especially to the heritage of the Greek church fathers, which has always been essential for the Greek church.[173]

On the other hand, the relationship between Greek Orthodoxy and Russia was not without its problems. There was a widespread conviction in seventeenth-century Moscow that the Greeks had lost their "ancient piety," and so they were sometimes asked to "correct their Christian faith" in one of the various monasteries. Greek merchants, and even sometimes also the clergy, were banned from entering Russian churches on account of their assumed proximity to the Turks.[174] Yet, there were also defenders of Greek culture among the Russian clergy. For example, Paisios Ligaridis, Metropolitan of Gaza, penned an epistle to Tsar Fedor Alekseevich in 1682 in which he praised Greek culture for its crucial role in forming Christianity.[175] The tendency to emphasize the Greek element on account of the Turkish influence on Greek culture grew in the following centuries.

Thus, the only theological treatise by Diamantis Rhyosis (ca. 1675–1746 or 1747), Korais' grandfather and teacher at the Patriarchal Academy, comprised 33 arguments against the Latin Church.[176] Latin scholastics were heavily criti-

171 Ibid., 336.
172 Ibid., 36–7.
173 Review by Theodor Nikolau, in *Orthodoxes Forum*, 4. Jg. 1990, Vol. 1+2: 110–4, here 112–3.
174 Olga B. Strakhov, *The Byzantine Culture in Moscovite Rus': the case of Evfimii Chudovskii (1620–1705)* (Cologne, Weimar and Vienna: Böhlau, 1998), 25.
175 Ibid., 35.
176 About Korais' roots v. Podskalsky, *Griechische Theologie*, op. cit., 330–1.

cized by the Greek church, and caused Eustratios Argentis to accuse the Western Church in the early eighteenth century of having made a grave mistake when Latin scholastics tried to merge Christian theology with Aristotle's thought, which meant that they had a number of misconceptions in their theology of the trinity. Argentis claimed that this was an aberrance of the theological approach initiated by the church fathers, who wanted to elevate philosophy to the level of theology.[177]

Individual scholars tried to reconcile the new ideas with the old. For example, the scholar Methodios Anthrakitis from Epirus (d. after 1736), who translated works of the French philosopher Nicolas de Malebranche (1638 – 1715), tried to establish a synthesis between Cartesian views and the illumination theology of St. Augustine. This angered the church, who accused him of being a heretic at the Patriarchal synod of 1723. He was allowed to resume teaching and priesthood two years later, but was obliged to swear loyalty to the Aristotelian philosophy as interpreted by Theophilos Korydalleas.[178] Some time afterwards, it was Patriarch Samuel who paved the way for a new phase of learning and translating.[179]

It was at this time that we find a first discussion of Descartes' epistemology and its opponents (Descartes; de Malebranche) in the theology of Vikentios Damodos (ca. 1700 – 1752), who lived in Venice and probably taught Eugenios Voulgaris. He also dedicated himself to the natural sciences.[180] Mt. Athos, with its long history of monasticism, also entered a new era when patriarch Kyrillos V founded in 1749/50 an academy, the so-called Athonias, which was located within the monastery of Vatopedi founded by Eugenios Voulgaris.[181] In general, proponents of the Greek Enlightenment came from the clergy.[182]

---

177 Ibid., 333 – 4; cf. Hans Freiherr von Campenhausen, *Griechische Kirchenväter* (Stuttgart, Berlin and Cologne: Kohlhammer, 1993), 119 – 20.

178 Podskalsky, *Griechische Theologie*, op. cit., 45.

179 Von Maurer, *Das griechische Volk*, Erster Band [1st vol.], op. cit., 429.

180 Podskalsky, *Griechische Theologie*, op. cit., 337, 341 – 2.

181 Martin Tamcke, *Das orthodoxe Christentum* (München: Beck, 2004), 66. Podskalsky, *Griechische Theologie*, op. cit., 59 – 60, cf.. Georgios Ainian, *Συλλογή ανεκδότων συγγραμμάτων του αιδίμου Ευγενίιου του Βουλγάρεως και τινων άλλων μετατυπωθέντων*, vol. I (Athens: K. Pallis, 1838), 54 – 64.

182 Max Demeter Peyfuss, *Die Druckerei von Moschopolis, 1731 – 1769: Buchdruck und Heiligenverehrung im Erzbistum Achrida* (Vienna et al.: Böhlau, 1989), 152 – 3; idem., "Die Akademie von Moschopolis und ihre Nachwirkungen im Geistesleben Südosteuropas," in *Wissenschaftspolitik in Mittel- und Osteuropa: Wissenschaftliche Gesellschaften, Akademien und Hochschulen im 18. und beginnenden 19. Jahrhundert*, ed. Erik Amburger, Michał Cieśla, and László Sziklay (Berlin: Ulrich Camen, 1976), 114 – 28, here 117 – 8.

The new learning came from the same spirit of reform that was sweeping through the entire Ottoman Empire, but it struggled against the more radical currents that bore an anti-clerical character. Associations to propagate the ideas of the French revolution had been founded in Smyrna by the end of the century. Symbols of the French Republic could often be seen on the streets of Aleppo and Constantinople.[183] Yet, Voulgaris' first attempt to open the Athos Academy for a new kind of learning was thwarted by quarrels within the clergy, since Greek monasticism had up until the eighteenth century been rather averse to education, and the era of interfaith dialogue ceased when the Eastern Roman Patriarchate was mostly hostile to all kinds of *kainotomia* ("inappropriate modernization").[184]

The Ionian Islands, which were ruled by Venice, were better connected to Western Europe than other parts of the Ottoman Empire, so that we find local libraries containing editions of Voltaire and the French Encyclopaedists at the end of the eighteenth century.[185] Scholars from the Ionian Islands played a crucial role in transmitting the ideas of an Enlightenment philosopher such as Christian Wolff (1679–1754), who gained some popularity in Ioannina and the Danube principalities. Although we know that his *Logik* was translated by Antonios Moschopoulos in 1785, there is still no bibliographical overview of Wolff's influence on modern Greek thought, which lasted for at least a century and stretched from roughly Paraskevas to Koumas.[186] Konstantinos Koumas (1770–1842), one of the main representatives of the German Enlightenment in Greece, was influenced by the German philosophers Immanuel Kant and Wilhelm Traugott Krug, who were read much less by Greek philosophers. Koumas was one of

183 Richard Clogg, "The 'Dhidhaskalia Patriki' (1798): An Orthodox Reaction to French Revolutionary Propaganda" in *Middle Eastern Studies*, Vol. 5, No. 2 (May, 1969): 87–115, here 87.

184 Podskalsky, *Griechische Theologie*, op. cit., 330; Ekaterini Mitsiou, "Interaktion zwischen Kaiser und Patriarch im Spiegel des Patriarchatsregisters von Konstantinopel," in *Zwei Sonnen am Goldenen Horn? Kaiserliche und patriarchale Macht im byzantinischen Mittelalter: Akten der internationalen Tagung vom 3. bis 5. November 2010*, ed. Michael Grünbart, Lutz Rickelt, and Martin Marko Vučeticć (Berlin: Lit, 2011), 79–96, here 85; cf. Martin Knapp, *Evjenios Vulgaris im Einfluss der Aufklärung: der Begriff der Toleranz bei Vulgaris und Voltaire* (Amsterdam: Hakkert, 1984), *passim*.

185 Clogg, "The ‚Dhidhaskalia Patriki'," op. cit., 89.

186 Panagiotis Noutsos, "Christian Wolff und die neugriechische Aufklärung," in *Evgenios Vulgaris und die neugriechische Aufklärung in Leipzig: Konferenz an der Universität Leipzig, 16.–18. Oktober 19965*, ed. Günther S. Henrich (Leipzig: Leipziger Universitätsverlag, 2003): 76–82, here 77, 70, 81.

the main proponents of the modern Greek Enlightenment; his works were printed in Vienna.[187]

Among the most important Greek philosophers to adopt the ideas of Wolff were Vikentios Damodos, Antonios Moschopoulos, and Eugenios Voulgaris. None knew Wolff personally, except Damianos Paraskevas from Sinope, who had studied in Frankfurt/Oder and Jena when the community of philosophers was divided between the followers of Thomasius on the one side and those of Wolff on the other. At that time, Paraskevas sympathized with the ideas of Wolff, whom he defended in two treatises. The focus of interest for all three Greek scholars was metaphysics.[188]

It was in Ioannina that the great reformist thinker and theologian Eugenios Voulgaris (1716–1806) worked as a teacher. It was here that he first came into contact with major figures of the Western Enlightenment such as Gottfried Wilhelm Leibniz (1646–1716), Wolff and John Locke, whose books he translated. After Ioannina, he went to Padua in Italy, where he studied philosophy, literature, physics and mathematics before becoming head of the educational center that had been founded by the brothers Marousos in Ioannina in 1740. He later appeared in Constantinople, where he taught at the Patriarchal Academy. Himself a modernizer of the traditional curriculum, Voulgaris had to leave Constantinople because of his critical stance towards Ottoman rule, and he went to the German city of Leipzig in around 1765. Also staying in Halle and Berlin, he spent seven years in Germany (i.e., the Holy Roman Empire).[189]

Voulgaris attempted in his *Logike* to revive the Eastern Roman body of philosophy through using Western rationalism, which is the reason that he broke with the old Aristotelian system.[190] He was not unambiguous about the ideas of Voltaire, whom he possibly met personally during his stay at Frederick's court, where Voltaire was a frequent guest. In any case, his attitude towards the French philosopher must have oscillated between "love and hate" (Podskal-

---

187 Kostas Th. Petsios, "Kants Kategorienlehre im Werk von Athanasios Psalidas," in *Evgenios Vulgaris und die neugriechische Aufklärung in Leipzig: Konferenz an der Universität Leipzig, 16.–18. Oktober 1996*, ed. Günther S. Henrich (Leipzig: Leipziger Universitätsverlag, 2003): 55–67, here 62 fn. 56. Nikos K. Psimmenos, "Ein philosophisches Vorlesungsheft aus dem philologischen Gymnasium zu Smyrna," in *Evgenios Vulgaris*, ed. Günther S. Henrich, op. cit.: 49–54, here 50–1.

188 Noutsos, "Christian Wolff und die neugriechische Aufklärung," op. cit., 78–9; Georgios Polioudakis, *Die Übersetzung deutscher Literatur ins Neugriechische vor der Griechischen Revolution von 1821* (Frankfurt/Main: Peter Lang, 2008), 61.

189 Von Maurer, *Das griechische Volk*, Erster Band [1st vol.], op. cit., 430; Polioudakis, *Die Übersetzung deutscher Literatur*, op. cit., 55–7.

190 Noutsos, "Christian Wolff und die neugriechische Aufklärung," op. cit., 80, 81–2.

sky).[191] His initial admiration for Voltaire faded over time and ended between 1766 and 1768. While he had praised him in his *Logike* (Λογική), he criticized him strongly two years later in his *Sketch on Religious Perfection* (Σχεδίασμα περὶ ἀνεξιθρησκείας) and even more in his commentary on his own translation of the *Essais sur les dissensions des Églises de Pologne.*[192]

Trained in rhetoric, logic, the natural sciences and metaphysics, Voulgaris was also influenced by the ideas of Locke, Leibniz and Wolff. He helped pave the way for educational books within the Greek Orthodox realm, and translated into Greek Wolff's *Die Elemente der Arithmetik und Geometrie.*[193] Himself a former student of Vikentios Damodos (d. 1752), his educational and intellectual efforts had an enormous impact on the modern history of Greek culture and society. Although he grew disillusioned when his ideas of reforming monasterial education fell on infertile ground among the mostly hesychastic monks, which caused him to refrain from further attempts at reform, his activities marked the beginning of a new epoch in modern Greek intellectuality.[194]

Leipzig is of special interest here. The city had connections both to Greece and to Russia since it was the Russian empress Catherine II who made Leipzig the favorite place of study for the young Russian nobility.[195] Voulgaris came into contact with Russia while he was still in Leipzig, which had an Orthodox community of both Greeks and Russians. After being invited to Russia, he was offered the post of archbishop in 1770, a post designed especially for him and was located in an area partly inhabited by Greeks.[196] He finally left for Russia in 1771, when Catherine II invited him with the view that he would support her reformist aspirations. He would spend the rest of his life there, dedicating himself to translating different scriptures into Greek, the most important appearing in St Petersburg in the year of his arrival. It was a Greek translation of Cath-

---

191 Podskalsky, *Griechische Theologie*, op. cit., 347.

192 Günther S. Henrich, "Als Denker Archaist, als Dichter auch Demotizist: Zu Vúlgaris' Paraphrase des Voltaireschen *Memnon*," in *Evgenios Vulgaris und die neugriechische Aufklärung in Leipzig: Konferenz an der Universität Leipzig, 16.–18. Oktober 1996*, ed. Günther S. Henrich (Leipzig: Leipziger Universitätsverlag, 2003), 99–111, here 101; cf. Konstantios Dimaras, *Η ιστορία της νεοελληνικής λογοτεχνίας* (Athens: Ermis, 1968), 133.

193 Podskalsky, *Griechische Theologie*, op. cit., 59–60, cf. G. Ainian, *Συλλογή ανεκδότων συγγραμμάτων*, op. cit., 54–64. In his *Metaphysics Voulgaris repeatedly refers to Descartes*, v. Evgenios Voulgaris, *Γενουηνσίου: Στοιχεία της Μεταφυσικής* (Vienna: Georgios Vendotis, 1806), 5. Θ'; Von Maurer, *Das griechische Volk*, Erster Band [1st vol.], op. cit., 430. On the translations v. Noutsos, "Christian Wolff und die neugriechische Aufklärung," op. cit., 80.

194 Podskalsky, *Griechische Theologie*, op. cit., 38–39, 42, 59–60.

195 Suppé, op. cit., 16.

196 Konstantinos Dimaras, *Νεοελληνικός Διαφωτισμός*, op. cit., 150–1.

erine's 1767 *Nakáz*, a template for a new constitution, which found a strong resonance in Voulgaris' thinking because it was influenced by ideas of the Enlightenment.[197]

Convinced that the Ottoman Empire was doomed, and possibly assuming that Western intervention would bring the Empire down, Voulgaris outlined his thoughts in a book with the title *Thoughts on the Current Crisis of the Ottoman State*, Voulgaris made clear in this book, which was published in St Petersburg and probably in 1772, that he saw the Empire as posing a threat to Europe that the European nations had to resist. France was the only country that he expected not to do so because it was so entangled commercially with the Ottoman Empire.[198]

Voulgaris possibly rubbed shoulders at Catherine's court with a great many Western scholars. After Catherine had acquired in 1765 the personal library of the French philosopher Denis who was Catherine's guest in 1773/4, when he outlined his ideas for reforming the Russian Empire on behalf of the empress.[199] Voulgaris, who later translated Herder into Greek, could possibly have learned about Herder through August Wilhelm Schlözer, was one of Herder's teachers, who was also at Catherine's court.[200] After Voulgaris became archbishop of the newly founded diocese of Kherson (Ukraine) in 1775, he dedicated himself to historical research. His interest should obviously be seen in the context of Catherine's "Greek Project," which was designed to make Russia the successor of the Eastern Roman Empire and the future ruler of Constantinople. The Black Sea was an area where Slavs from the Balkans and Greeks had settled in newly

197 Von Maurer, *Das griechische Volk*, Erster Band [1st vol.], op. cit., 430; Vasilios N. Makrides, "Orthodoxie und Politik: die russisch-griechischen Beziehungen zur Zeit Katharinas II.," in *Katharina II., Rußland und Europa: Beiträge zur internationalen Forschung*, ed. Claus Scharf (Mainz: Zabern, 2001): 85–119, here 103; Paschalis Kitromilides, "Η πολιτική σκέψη του Ευγενίου Βούλγαρη," in *Evgenios Vulgaris und die neugriechische Aufklärung in Leipzig: Konferenz an der Universität Leipzig, 16.–18. Oktober 1996*, ed. Günther S. Henrich (Leipzig: Leipziger Universitätsverlag, 2003), 85–98, here 97. Also the Scottish engineer James Watt decline an invitation by Catherine, v. Linda Colley, *Britons: Forging the Nation 1707–1837* (New Haven and London: Yale University Press, 2009), 125.

198 Makrides, "Orthodoxie und Politik," op. cit., 103.

199 Denis Diderot, "Observations sur le Nakaz" (1774), in idem, *Oeuvres politiques*, ed. Paul Vernière (Paris: Garnier Frères, 1963), 350; Andreas Heyer, *Materialien zum politischen Denken Diderots: Eine Werksmonographie* (Hamburg: Kovač, 2004), 432. Peter Pappageorg, "Einiges über die griechische Gemeinde in Leipzig," in *Hellas-Jahrbuch* 1930: 123–5; Suppé, op. cit., 15 Fn. 6.

200 Günter Mühlpfordt, "Hellas als Wegweiser zur Demokratie. Griechenmodell und Griechenkritik radikaler Aufklärer–Antikerezeption im Dienst bürgerlicher Umgestaltung," in *Griechenland–Byzanz–Europa: Ein Studienband*, ed. Joachim Herrmann, Helga Köpstein and Reimar Müller (Amsterdam: Gieben, 1988), 225–69, here 260.

founded towns that were given Greek names.[201] Voulgaris, who created the word *politokratía* (i.e., republic), championed the autonomous, non-monarchical state that came into being when the Ionian Islands fell to Russian-Turkish control between 1799 and 1806.[202] Voulgaris himself had an enormous impact on later generations, although his own aspirations for reform were rather moderate and, as a cleric, he never abandoned a somewhat traditionalist thinking. We should not forget that the Greek church has always objected to the secularist thought of the French Enlightenment.[203] In any case, Voulgaris is still remembered as the "great teacher of the revolutionary generation" (Podskalsky),[204] since he paved the way for a cultural revival that peaked with the ideas of Adamantios Korais.[205]

Athanasios Psalidas (1767–1829), one of the chief proponents of the modern Greek Enlightenment who had studied in Russia and Vienna, walked in Voulgaris' footsteps. Although a defender of religion, Psalidas was far from opposed to the Enlightenment; he was a follower of Kant and deemed it impossible for the human to realize and verify the existence of God. His thoughts on the freedom of man are influenced by Kant, and his mention of Kant in his treatise is the first mention in Greek literature. The curriculum that Psalidas designed for the school that he ran in Ioannina was influenced by the Enlightenment.[206]

Psalidas often referred to Locke, Spinoza, Helvétius, Hobbes, Leibniz, Wolff, and Kant. In his book *True Well-Being* (Ἀληθὴς Εὐδαιμονία),[207] Psalidas established four principles to realize human well-being: the existence of God, the immortality of the soul, reward after death, and human freedom. Although Psalidas held reason in high esteem, he considered it unable to prove the existence of the

---

201 Polioudakis, *Die Übersetzung deutscher Literatur*, op. cit., 58; Dan Diner, "Zweierlei Osten," op. cit., 107–8.

202 Kitromilides, "Η πολιτική σκέψη του Ευγενίου Βούλγαρη," op. cit., 97; Hamish Scott, *The Birth of a Great Power System, 1740–1815* (Harlow and New York: Pearson/Longman, 2006), 323; Alfred J. Rieber, *The Struggle for the Eurasian Borderlands: From the Rise of Early Modern Empires to the End of the First World War* (Cambridge and New York: Cambridge University Press, 2014), 310.

203 Albert Hourani, "Kultur und Wandel: Der Nahe und Mittlere Osten im 18. Jahrhundert," in idem, *Der Islam im europäischen Denken* (Frankfurt/ Main: S. Fischer, 1994), 169–201, here 181–2.

204 Podskalsky, *Griechische Theologie*, op. cit., 344–353. Papoulia in her book *Από την αυτοκρατορία στο έθνος-κράτος* (op. cit.) does not mention Voulgaris at all. More information is offered by Dimaras, *Νεοελληνικός Διαφωτισμός*, op. cit., *passim*.

205 Von Maurer, *Das griechische Volk*, Erster Band [1st vol.], op. cit., 431.

206 Dimaras, *Νεοελληνικός Διαφωτισμός*, op. cit., 155–6; Petsios, "Kants Kategorienlehre," op. cit., 55–7, 63.

207 Complete title: *Ἀληθὴς Εὐδαιμονία ἤτοι βάσις πάσις θρησκείας: Vera Felicitas sive Fundamentum omnis Religionis, graece et latine*, vol. 1 (Vienna: Baumeister, 1791).

four principles. Voulgaris criticized his argumentation.[208] Kant's ideas began to spread in Greece between 1795 and 1820, shortly after Psalidas' work was published, and they led to heated discussion. Other than that, Psalidas was well-versed in many disciplines and had a strong interest in natural law.[209]

The idea of natural law did in fact seem compatible with the teachings of Christianity, since the church had always differentiated between divine and human law before medieval canon law gave birth to the modern idea of human rights. Natural law gained in status in the Latin West even up until the law of the church was judged in accordance with it.[210] Voltaire played a pivotal role in the Enlightenment discourse. Nikiforos Theotokis (1731–1800), another cleric and bearer of the Enlightenment, also contributed to spreading the ideas of Voltaire. Following Voulgaris' lead, he aimed to improve the level of education and published a book each on geography, physics, and mathematics.[211]

Even more important was Christodoulos Pamplekis, who was one of Voulgaris' students at the Athos Academy before he established contact with philosophers in Germany and France, and worked as a lecturer in Vienna and Leipzig. He is mainly known for his account of the history of philosophy spanning from the antiquity to modern France and Germany. He published his *Trophies of the Orthodox Faith* (Τρόπαιον τῆς ὀρθοδόξου πίστεως) in 1791, where he makes frequent reference to Voltaire. Pamplekis' claim that God's existence had been a human idea since the beginnings of humanity was strongly criticized by the church, which accused him of being obsessed by the ideas of the already condemned Voltaire, as well as by those of Rousseau and Spinoza. Two years later, in 1793, the Eastern Church condemned his writings in an encyclical, and he was excommunicated after he refused to withdraw his claim.[212]

However, the new learning spread in different centers and waves. One of these was the so-called "Renaissance of Moschopolis," a movement of new learning in eighteenth-century Epirus and Macedonia that was begun by Theodoros Anastasios Kavalliotis, who is thought to have been a student of Voulgaris'. Moschopolis hosted the "New Academy" (Νέα Ακαδημία) in the mid-eighteenth century, at a time when the old Aristotelian thinking was in decline and was giving way to a new "critical thinking." This academy ran a curriculum similar to those at institutions in Western Europe. The academy contributed in the South-

208 Petsios, "Kants Kategorienlehre," op. cit., 56–8.

209 Ibid., 62–3.

210 Rhonheimer, *Christentum und säkularer Staat*, op. cit., 89–90.

211 Mackridge, *Language and National Identity*, op. cit., 89–90; Von Maurer, *Das griechische Volk*, Erster Band [1st vol.], op. cit., 430.

212 Suppé, op. cit., 36–7; Podskalsky, *Griechische Theologie*, op. cit., 372–6.

east Europe of the late nineteenth century to the cultural revival not only of the Greeks but of also other nations. Under Sevastos Leontiadis (d. 1765), the "New Academy" would be a hotspot of reformist thinking that was critical of the tradition.[213]

Kavalliotis began to teach in Moschopolis not later than in 1743. He stressed the need to study the scriptures of pagan Greek antiquity. Kavalliotis seemed to have taken a rather moderate stance between the neo-Aristotelians and the neo-Platonists, and he is therefore now usually regarded as having been part of the Enlightenment. Law constitutes in Aristotelian ethics a higher force or principle that holds individual interests in society together. Leontiadis stood in this controversy against his Aristotelian opponents, while championing a rather critical way of thinking. His philosophy was influenced by Voulgaris and drew on Enlightenment thinkers such as Leibniz, Wolff, de Malebranche (1638–1715), and Descartes (1596–1650). Kavalliotis' successor as head of the New Academy was the Swedish scholar Johann Thunmann, who entered office in 1750 and became well-known for his research on contemporary Albanian.[214]

The Albanian elite found its center of learning in the Zosimaia School in Ioannina.[215] Similar schools were established in the provincial towns of the Ottoman Empire, wherever Greeks lived. This happened initially without the support of the Sublime Porte and partly against its will. But, later, under Selim III (1789–1807), the government approved of and even sustained such schools. The school in Bucharest headed by the popular Lambros Fotiadis, who was followed by the even more popular Neofitos Doukas and the latter's friend Stefanos Komitas, became an important hub of knowledge. The school on the island of Chios, where the philhellene Julius David taught French and Latin, had no fewer than fourteen teachers.[216]

Of course, not all of Voulgaris' students advocated the French ideas of the Enlightenment. Athanasios Parios, a former student of Voulgaris', became a fervent opponent of innovations in the natural sciences and philosophy, and he op-

213 Papoulia, *Από την αυτοκρατορία στο εθνηκό κράτος*, op. cit., 274. Peyfuss, *Die Druckerei von Moschopolis*, op. cit., 152; idem, "Die Akademie von Moschopolis," op. cit., 117–8, 123.

214 Peyfuss, *Die Druckerei von Moschopolis*, op. cit, 152–3; idem, "Die Akademie von Moschopolis," op. cit., 117–8, 123. Descartes also had a strong influence on the reformist thinker Iosepos Moissiodax (Ιώσηπος Μοϊσιόδαξ, 1725–1800) who propagated mathematics as being the foundation of all understanding. Iosepos Moissiodax, v. Raphael Demos, "The Neo-Hellenic Enlightenment (1750–1821)," in *Journal of the History of Ideas*, Vol. 19, No. 4 (1958): 523–41, here 536. Guariglia, op. cit., 92.

215 Papoulia, *Από την αυτοκρατορία στο εθνηκό κράτος*, op. cit., 274.

216 Von Maurer, *Das griechische Volk*, Erster Band [1st vol.], op. cit., 434–6.

posed any revolutionary thought of French origin. For many years the principal of the academy at Chios, he became head of the Athos Academy in 1771, and later on of the Patriarchal Academy.[217] Athanasios was not the only cleric to remain loyal to Ottoman rule. When Nikodimos Agioritis (d. 1809) defended the political *status quo*, Orthodoxy in general became increasingly hostile to the ideas of the French Enlightenment.[218]

Similarly, Makarios, metropolitan of Corinth (d. 1805) and a former student of Parios', defended the Orthodox faith against the ideas of the Enlightenment.[219] In a treatise called *Fatherly Teaching* (Διδασκαλία Πατρική), its author, probably Anthimos of Jerusalem (1717–1808), justified why Greek subjects should remain under Ottoman rule. Anthimos had maintained good relations with sultan Selim III (1789–1807), who was cooperative in granting Orthodox rights in the Holy Land. Born in 1717, Anthimos became Patriarch of Jerusalem in 1788 and stayed in office until his death in 1808. His theological concept of "voluntary slavery" (*εθελοδουλία*) legitimized servitude to Turkish rule.[220] Meanwhile, Napoleon turned a new page in history when he invaded Egypt.[221]

1.5 Napoleon in Egypt

The decision of the Maronite Synod of 1736 to found a modern educational system began a new wave of learning in the Levantine Arab world, too, with this world becoming later on a means of transmitting ideas from Western Europe to the Middle East.[222] It welcomed Christian Arab intellectuals who had often been born into rising merchant families with connections to other continents, and who had a command of more than one language.[223] Although wealthy, the merchant class never managed to shape society as its counterpart in the West did, which was mainly due to economic insecurity. Also, as Clark (2014) has put it, the practice in Islamic societies "of imposing taxes on religious minorities

---

217 Podskalsky, *Griechische Theologie*, op. cit., 40, 359, 360 (fn. 1513), 361; Clogg, "The ‚Dhidhaskalia Patriki'," op. cit., 95.
218 Podskalsky, ibid.
219 Ibid.
220 Clogg, "The ‚Dhidhaskalia Patriki'," op. cit., 95–6, 98, cf. 102.
221 Hösch, *Geschichte der Balkanländer*, op. cit., 145.
222 Theodor Hanf, "Die christlichen Gemeinschaften im gesellschaftlichen Wandel des arabischen Vorderen Orients," in *Orient* 1 (1981): 29–49, here 34–5.
223 Hisham Sharabi, *al-muthaqqafūn al-ʿarab wa-l-gharb: ʿaṣr an-nahḍa 1875–1914* (Beirut: Dar an-Nahar li-n-Nashr, [1978] 1991), 61; Gunnar Hering, *Ökumenisches Patriarchat*, op. cit., 147–8.

tended to recruit to Islam the lowest socioeconomic strata of conquered societies."[224]

However, the merchant class was under the influence of two main intellectual currents: the Enlightenment on the one side and nineteenth-century positivism and liberalism on the other. Montesquieu was of special interest for Arab proponents of Enlightenment thinking because he justified arguing for a separation of powers, and Rousseau's discussion of the "general will" (*volonté générale*) was discussed in numerous articles in journals such as *al-Muqtaṭaf* and *al-Hilāl*.[225] These ideas were not completely new to Arab and Middle Eastern reformists, who could connect them to endogenous developments such as the so-called ego-documents. These documents appeared in the seventeenth century and promoted the idea of subjectivity and individuality. The first was written by Ḥāfiẓaddīn al-Qudsī (died in 1645/6), and similar autobiographical writings in the Ottoman language began to follow in the second half of the seventeenth century.[226]

Napoleon conquered Egypt for primarily strategic reasons, but the conquest conjured up age-old memories since Egyptians represent in the ancient Greek tradition worldly wisdom. This is why philosophers used to defend Egypt against accusations of Bible followers that they followed the most abominable form of idolatry based on animal-like juggernauts.[227] There was renewed European interest in the late sixteenth century, which saw a wave of Western European scholars travelling to Egypt and was followed by another wave in around 1740, when two scholars, Richard Pococke from England and Frederik Norden from Denmark, reached Upper Egypt at the same time. Both were driven by the desire to explore the glory of ancient Egypt. Almost sixty years later, Egypt was conquered by Napoleon.[228]

---

224 Bernard Lewis, *Faith and Power: Religion and Politics in the Middle East* (Oxford: Oxford University Press, 2010), 68–9. Gergory Clark, *The Son Also Rises: Surnames and the History of Social Mobility* (Princeton und Oxford: Princeton University Press, 2014), 238–9.

225 Hourani, "Kultur und Wandel," op. cit., 181–2.

226 Elger, "Selbstdarstellungen aus Bilâd ash-Shâm," op. cit., 130–1; cf. Ulrich Haarmann, "Ideology and history, identity and alterity: The Arab image of the Turk from the 'Abbasids to modern Egypt," in *International Journal of Middle East Studies*, Vol. 20 (1988): 175–96, here 182, 187 and *passim*. Albert Hourani, *Arabic Thought in the Liberal Age: 1798–1939* (Cambridge: Cambridge University Press, [12th ed.] 2002), 278.

227 Hans Blumenberg, *Das Lachen der Thrakerin: Eine Urgeschichte der Theorie* (Frankfurt/Main: Suhrkamp, 1987), 49–50.

228 Osterhammel, *Die Entzauberung Asiens, op. cit.*, 105; cf. Peter J. Ucko and T. C. Champion, *The Wisdom of Egypt: Changing Visions Through the Ages* (London: UCL Press, 2003), 149.

Egypt was of greater interest to the Ottoman Empire, which is the reason that Napoleon decided to use it as his launch pad to attack British rule in India and to cut the British supply route.[229] This was probably triggered by British attempts to curb the impact of the French revolution.[230] Napoleon himself spoke publicly of the genius of liberty that he was determined to spread to the farthest lands.[231] However, he was not the first to envision such a scheme. Leibniz had already presented a similar project to Louis XIV.[232] As Blumenberg pointed out, occupying Egypt meant connecting with its ancient spirit.[233] When Napoleon conquered Egypt in July 1798 with 40,000 soldiers under his command, Mamlūk rule was defeated within a month, its defeat culminating in a battle near the pyramids.[234] Napoleon proclaimed in his address to the victorious army in June 1798:

> [...] the Mamluk Beys, who fostered nothing but English commerce and extorted our own merchants in a shameful way while tyrannizing the wretched inhabitants of the Nile shores, will no longer exist a few days after our arrival.
>
> The peoples with whom we will live now are Mohammedans; their first article of creed says: "There is no other God but God and Mohammed is His Prophet."
>
> Don't answer them back; treat them like we treated the Jews and the Italians; treat the muftis and imams with the same respect you have showed for rabbis and bishops.
>
> Be tolerant with the ceremonies prescribed by the Quran and with the mosques in the same way you were tolerant with the monasteries, the synagogues and the religion of Moses and Jesus Christ.
>
> The Roman legions used to protect all of the religions. Here you will find other customs than those in Europe: You have to get used to them ...

---

229 Carl Brockelmann, *Geschichte der islamischen Völker und Staaten* (Munich and Berlin [2nd ed.] 1943, Reprint Hildesheim and New York: Olms, 1977), 313.

230 Juan Cole, *Napoleon's Egypt: Invading the Middle East* (New York et al.: Palgrave Macmillan, 2008), 3.

231 Ibid., 5.

232 Werner Schneiders, "Deus subjectum. Zur Entwicklung der Leibnizschen Metaphysik," Originally published in *Leibniz à Paris (1672–1676)*, Vol. II, *La philosophie de Leibniz, Studia Leibnitiana, Supplementa*, ed. K. Müller, H. Schepers and W. Totok, Vol. XVIII (Wiesbaden: Steiner, 1978), 21–31, repr. in *Werner Schneiders: Philosophie der Aufklärung – Aufklärung der Philosophie: Gesammelte Studien, zu seinem 70. Geburtstag*, ed. Frank Grunert (Berlin: Duncker & Humblot, 2005): 13–24, here 22; idem., "Harmonia universalis. Harmonie als Schlüsselbegriff der Leibnizschen Philosophie," originally published in *Studia Leibnitiana*, Vol. XVI/1 (1984),: 27–44, reprinted in op. cit.: 25–47, here 25.

233 Hans Blumenberg, *Präfiguration: Arbeit am politischen Mythos*, ed. Angus Nicholls and Felix Heidenreich (Berlin: Suhrkamp, 2014), 23–4.

234 John Darwin, *Der Imperiale Traum: Die Globalgeschichte großer Reiche 1400–2000* (Frankfurt and New York: Campus, 2010), 178.

> The first city we will enter was built by Alexander. Every step we take will prompt the French to competitive zeal.[235]

In order to legitimize his conquest and to make people obey the new rule, Napoleon sought the approval of the clerics at Azhar University. He was informed by sheikh ʿAbdullāh ash-Sharqāwī that, if he wanted the Muslim to follow his banner, he had to convert to Islam. He would then be able simply to conquer the Orient and to restore the glory of the Prophet's fatherland.[236] Although Napoleon's army did not have any particular religion, it was full of believers, many of whom felt alienated by the anticlericalism of the French Revolution, and this made them susceptible to the religion of Islam. One of the converts was the commander of Rosetta, Jacques-François Menou, a long-term security officer.[237] Napoleon's conquest was an attack on the British colonial empire in India, and Britain therefore tried to encourage the Ottomans to resist Napoleon. This French-British conflict saw both sides claiming to be the true protectors of Islam.[238]

As Kohn (1929) noted, Napoleon's aspirations brought Egypt "into the glare of modern European politics" for the first time.[239] He also tried to legitimize his rule by suggesting that French deism had commonalities with Islam. He made clear in his original Arabic declaration that the French army was devoid of any particular religion, and even rejected the dogma of the trinity, declaring that his army was a "muslim" (with a small *m*) army. Although this clearly served his political aims, Napoleon's interest in Muhammad might possibly have been genuine.[240]

Appropriately, Napoleon's entourage included 130 scholars who were members of the *Commission des Sciences et des Arts*. Some of these scholars would be employed by the *Institut d'Égypte*, where they drafted a comprehensive report on the time of occupation up until 1801. This report, titled *Description de l'Égypte*,[241]

---

235 Napoleon Bonaparte, *Napoleons Briefe*, ed. Friedrich Schulze (Leipzig: Insel-Verlag, 1912), 98.

236 Cole, *Napoleon's Egypt*, op. cit., 127–8.

237 Ibid., 134–5.

238 Faisal Devji, "Islam and British Imperial Thought," in *Islam and the European Empires*, ed. David Motadel (Oxford: Oxford University Press, 2016): 254–68, here 257; Devji follows the argumentations of Cole in *Napoleon's Egypt*, op. cit., chap. 7.

239 Hans Kohn, *A History of Nationalism in the East* (New York: Harcourt, Brace and Co., 1929), 319.

240 Cole, *Napoleon's Egypt*, op. cit., 129.

241 Complete title: *Description de l'Égypte ou receuil des observations et des recherches qui ont été faites en Égypte pendant l'expédition de l'armée francaise, publié par les ordres de Sa Majesté l'Empereur Napoleón le Grand.*

also described a future canal to be built through the Isthmus of Suez. Because of its detailed information, it inspired thousands of European scholars, engineers and technicians to work in Egypt, one result of which was the upgrading of the irrigation system. As a consequence, a business community became increasingly interested in French commercial ties with Egypt.[242]

The Ecumenical Patriarch Gregory V (1746–1821) dismissed the "foolish wisdom of Europe," (cf. 1. Cor 3:19) and called for loyalty to the sultan even after the campaign in Egypt. He feared the wave of atheism more than Muslim rule.[243] The views of the Orthodox hierarchy were in line with those of the Porte, which no longer tolerated the spread of revolutionary French ideas on its territory. The first cracks in the Ottoman Empire began to appear with the siege of Malta in 1798 and the expedition to Egypt in the same year, and the Greeks felt encouraged to pursue their own cause and hoped for French support.[244]

Overall, Napoleon had a negative image in Russia, where he appeared as a kind of Russian antagonist fanning Russian nationalism.[245] At the same time, German Enlightenment figures depicted him as a new Xerxes who was being fought by nations that were following in the footsteps taken by the Greeks against the Persians.[246] Nietzsche would therefore call Napoleon a "synthesis of prehistoric man and superman."[247]

Although there is no reason to believe that his conquest of Egypt was welcomed in the Islamic world, the claim made by Cole (2008) that Napoleon was aware that every educated Egyptian Muslim would compare the French conquest to the crusades is doubtful, since the crusades had never played an important role in the collective Arab memory, something that would not change until the late nineteenth century.[248] Napoleon styled his army as a kind of freemason group with himself as Grand Master.[249] Or he saw his expedition to Egypt as reviving the endeavour made by Alexander. Having toyed even with the idea of de-

---

242 Antoine B. Zahlan, "The Impact of Technology Change on the Nineteenth-Century Arab World," in *Between the State and Islam*, ed. Charles E. Butterworth and I. William Zartman (Cambridge and Washington: Woodrow Wilson Center Press, 2001): 31–58, here 33–4.

243 Clogg, "The 'Dhidhaskalia Patriki'," op. cit., 2, 94.

244 Ibid., 92–3.

245 Ibid.

246 Mühlpfordt, "Hellas als Wegweiser zur Demokratie," op. cit., 251.

247 Friedrich Nietzsche, "Zur Genealogie der Moral," in idem, *Sämtliche Werke: kritische Studienausgabe in 15 Bänden*, ed. Giorgio Colli und Mazzino Montinari, vol. 5: *Jenseits von Gut und Böse; Zur Genealogie der Moral* (Munich: DTV and De Gruyter, [1988/1999] 2012): 245–412, here 288.

248 Cole, *Napoleon's Egypt*, op. cit., 127.

249 Ibid., 5.

claring himself the son of God, Napoleon finally refrained from doing so for practical reasons. In any case, the local population was muted in its response, and he therefore gave up on the idea of transplanting the Enlightenment to Egypt. Designed to connect France with the ancient culture of Egypt, and thereby to legitimize French rationalism, the expedition turned out to be a failure. As Blumenberg (1996) pointed out, Napoleon could not bear the idea that the Orient was more different than he had expected and could therefore not be used to support his own theophany.[250]

Napoleon never reached India, but planned for a campaign into Syria. However, Nelson destroyed the French fleet at Abukir and cut Napoleon's supply line to France. Napoleon's campaign to Syria was bound to fail when uprisings occurred in Egypt and the Ottomans declared war on France. Napoleon was forced to flee clandestinely to France in August 1799. British troops sent from the British islands and from India conquered Cairo in June 1801 and brought the French project in the Middle East to an end.[251] Napoleon nonetheless had a lasting impact, at least in Europe, where he spread the ideas of the French Revolution and thus laid the ground for inclusive institutions and economic prosperity – a major difference to the Middle East.[252]

It was the writings of Voltaire that found great interest and even affirmation in the Christian Arab context. This parallels those Greek intellectuals during the war of liberty who distanced themselves from the older generation of clerics and their loyalty to the Ottoman government, but embraced French criticism of the church and sought independence from the Ottoman Empire.[253] As Sharābī (1970) noted, not only was the Eastern Church strongly criticized by intellectuals such as Shidyāq; this criticism was even articulated in typical Voltairian language.[254]

---

250 Blumenberg, *Arbeit am Mythos*, op. cit., 55.

251 Darwin, *Der Imperiale Traum*, op. cit., 178.

252 "The French Revolution thus prepared not only France but much of the rest of Europe for inclusive institutions and the economic growth that these would spur," Daron Acemoglu and James A. Robinson, *Why Nations Fail*, op cit. 291.

253 Podskalsky, *Griechische Theologie*, op. cit., 385; Dimaras, *Νεοελληνικός Διαφωτισμός*, op. cit., 151; cf. Kreutz, *Das Ende des levantinischen Zeitalters*, op. cit., 45–6.

254 Hisham Sharabi, *Arab Intellectuals and the West: The Formative Years, 1875–1914* (Baltimore: John Hopkins Press, 1970), 67–8.

2 The Mediterranean Dawn

2.1 The Greek Rise to Independence

The Greek desire for independence might well date back to the early eighteenth century, when, as Maurer was convinced, Peter the Great stood at the Prut.[1] Even before, in the first half of the sixteenth century, the region had been rocked by insurgency when the fleets of Charles V, king of the Holy Roman Empire, and sultan Süleyman fought in the Eastern Mediterranean, bringing mayhem to the population of each wherever they disembarked. When Charles' fleet landed on the Peloponnese in order to conquer the city of Methoni and its garrison in 1532, the local Christian population of Koroni felt encouraged to rise up against and expel the Ottomans from their town. The Ottomans struck back two years later, and Greek and Albanian insurgents finally had to flee to Naples, where Charles gave them refuge.[2] Uprisings against the Ottomans also occurred in 1769 and 1770, both incited by the Russian empress Catherine II, but both were suppressed.[3] Yet, the seeds of independence had been sown, and the efforts to gain independence became more powerful when Rigas Ferraios Velestinlis (ca. 1757–98) merged them with the ideas of the French Revolution.[4]

The protégé of Dimitrios Katartzis (1730–1807), and a Phanariot, Enlightenment figure and encyclopaedist, Rigas propagated the idea of an independent Greek state through poems and songs.[5] While in Vienna, which was an important hub of ideas stemming from the West and which was connected to the Southeast of the continent, Rigas prepared a map in 1798 that showed the Balkans and the shores of the Eastern Mediterranean.[6] This made him the progenitor of a Greek nation-state based on a revived ancient-style democracy as well as on the ideas of the French Revolution, a nation-state that stretched far beyond its cur-

1 Von Maurer, *Das griechische Volk*, Erster Band [1st vol.], op. cit., 28.
2 Peter A. Mazur, "A Mediterranean Port in the Confessional Age: Religious Minorities in Early Modern Naples," in *A Companion to Early Modern Naples*, ed. Tommaso Astarita (Leiden et al.: Brill, 2013): 235–56, here 227; Charles A. Frazee, *Catholics and Sultans: The Church and the Ottoman Empire 1453–1923* (Cambridge et al.: Cambridge University Press, 2006), 39.
3 Von Maurer, *Das griechische Volk*, Erster Band [1st vol.], op. cit., 74.
4 Hösch, *Geschichte der Balkanländer*, op. cit., 122, 167.
5 Ibid., 145. Dimaras, *Νεοελληνικός Διαφωτισμός*, op. cit., 257–8.
6 Stephen G. Xydis, "Modern Greek Nationalism," in *Nationalism in Eastern Europe*, ed. Peter F. Sugar and Ivo John Lederer (Seattle: University of Washington Press, 1994): 207–58, here 228. On the significance of Vienna for the Slavic Enlightenment in Southeast Europev. Matl, op. cit., 120, 125, esp. 446–7, 449–50.

https://doi.org/10.1515/9783110634006-004

rent borders. He was therefore also the progenitor of what would later be called the "great idea," i. e., the idea of a Greek imperial state that would succeed the Ottoman Empire.[7] Rigas Velestinlis was finally taken into custody in Trieste in December 1797, while carrying a chest of flyers propagating the ideas of the French Revolution and calling for the end of Ottoman tyranny. The Austrian authorities handed him over to the Ottomans, who executed him in Belgrade in 1798.[8]

According to the records of the interrogation, the forty-year-old Rigas admitted to having performed, by singing and playing the flute, a Greek poem, the "Thourios Hymnos" (Song of War), which begins with the stanza "How long, you comrades?" (*ώς πότε παλικάρια*) and which he used to called for revolt against the Turks. He was also accused of having issued 1,200 copies of a portrait of Alexander the Great for circulation among the Greek community. This, by the way, reflected a popular tendency within the Eastern Mediterranean space, which saw Alexander being used for all kinds of political agendas. By the time he was arrested, only a few copies had been distributed. Rigas confessed to his schemes for liberating Greece from the Turkish yoke and said that he would rather live under the devil's rule than under that of the Turkish sultan.[9]

Himself a constitutionalist, Rigas also toyed with the idea of translating into Greek the French constitution and Montesquieu's "Spirit of the Law" (*Esprit des lois*). His own vision comprised a Greek hegemonic state, which ignored the Slavic peasants, who identified Greekdom with exploitation, which would make them vulnerable to Bulgarian nationalism.[10] On the other hand, Rigas' concept of democracy was modelled on both the antiquity and on modern thinkers such as Rousseau.[11]

---

7 Papoulia, *Από την αυτοκρατορία στο εθνηκό κράτος*, op. cit., 176 – 7; Vakalopoulos, *Griechische Geschichte*, op. cit., 85; Hösch, *Geschichte der Balkanländer*, op. cit., 211.

8 Hösch, *Geschichte der Balkanländer*, op. cit., 145; Clogg, "The ‚Dhidhaskalia Patriki'," op. cit., 90.

9 From the interrogation records with Rigas Velestinlis, Vienna, April 3, 1798, in *Documents inédits concernant Rigas Vélestinlis et ses compagnons de Martyre*, ed. Émile Legrand (Paris: E. Leroux, 1892), 58 – 60, 62, 64, 66.

10 From the interrogation records with Rigas Velestinlis, Vienna, April 3, 1798, in *Documents inédits*, op. cit., 68; Xydis, "Modern Greek Nationalism," op. cit., 228. Papoulia, *Από την αυτοκρατορία στο εθνηκό κράτος*, op. cit., 178 – 9. About Rigas' life cf. Christopher M. Woodhouse, *Rhigas Velestinlis: The Proto-Martyr of the Greek Revolution* (Limni Evia: D. Harvey, 1995), *passim*. Basil C. Gounaris, "Social cleavages and national ‚awakening' in Ottoman Macedonia," in *East European Quarterly* XXIX, No. 4 (1996): 409 – 26., here 416.

11 Papoulia, *Από την αυτοκρατορία στο εθνηκό κράτος*, op. cit., 176 – 7; Vakalopoulos, *Griechische Geschichte*, op. cit., 85.

Greece also saw in the early nineteenth century the birth of a new kind of literature, the so-called *mismagies*, which were collections of folk songs that, conveying motifs of love, passion and desire, spread ideas of individuality. The *mismagies* found their way into the nationalist literature that glorified the Greek aspirations for independence. Some of the *mismagies* also found their way into short stories of French origin, as was the case in the writings of Rigas Ferraios Velestinlis. His translations deserve a closer look, since he tried to popularize the ideas of the French Revolution by means of literature. For example, he translated into Greek Nicolas Restif de la Bretonne's *School of Delicate Lovers*, a collection of somewhat eroticized and loosely bound episodes, and deployed them with those *mismagies* that had become popular among resistance fighters (i. e., the *klephtes*). Rigas' method of blending literary work with folk elements echoed the position of the radical French Enlightenment, which went to great lengths to deny all claims of authorship and praised plagiarism as progressiveness.[12]

The *mismagies*, with their focus on issues of love and romance, became popular at a time when European societies, and especially France and Britain, were developing a more individualistic trait that spilled over into literature and promoted the new genres of love letters and novels.[13] What also helped the *mismagies* was that the Greeks of that time were widely believed to be identical to the ancient Greeks.[14] When Rigas was executed in 1798, he left a mark on how the occupied Greeks were viewed and evoked a new wave of philhellenism in Europe,[15] which critics such as Fallmerayer and Dragoumis would later ridicule.

At that time, Austria was facing war on two fronts: with Prussia, and with the Ottoman Empire, which, in opposing the Greek national movement, was threatening to shake the Austrian position on the Balkans to its core. The records state in a rather conciliatory tone that Rigas sympathized with the "democratic catechism," and that he was against not only the Turkish rule but also tyranny in

---

12 Michael Kreutz, ‚Empire and Enlightenment: Greek Poetry in Ottoman Letters," in *Marginal Perspectives on Early Modern Ottoman Culture: Missionaries, Travellers, Booksellers*, ed. Ralf Elger and Ute Pietruschka [= *Hallesche Beiträge zur Orientwissenschaft*, Vol. 23/2013] (Halle (Saale): ZRS, 2013): 1–15, here 6–8.; Von Maurer, *Das griechische Volk*, Erster Band [1st vol.], op. cit., 46, cf. Bd. II, 24.

13 Lawrence Stone, "Der Wandel der Werte in England 1660 bis 1770: Säkularismus, Rationalismus und Individualismus," in *Kulturen der Achsenzeit*, Bd. II: *Ihre institutionelle und kulturelle Dynamik*, Teil 3: *Buddhismus, Islam, Altägypten, westliche Kultur*, ed. Shmuel N. Eisenstadt (Frankfurt/Main: Suhrkamp, 1992): 341–57, here 348–9.

14 Georg Ludwig von Maurer, *Das griechische Volk, op. cit.*, Zweiter Band [2nd vol.], 35–6.

15 Von Maurer, *Das griechische Volk*, Erster Band [1st vol.], op. cit., 439–40.

general.[16] Rumours circulated in 1797/98 that a revolution threatening to rip through the entire Ottoman Empire was imminent.[17] Ideas of Greek independence soon became manifest in organizations set up to that end and especially in the "Society of Friends" (*filiki etairia*), which was founded in Odessa in 1814 by three merchants, Nikolaos Skoufas, Athanasios Tsakaloff and Emmanouil Xanthos, and which aimed to prepare the Greeks for the revolt to come.[18]

The Greeks acquired their first taste of independence when the Ottoman-Russian alliance granted autonomy to the Ionian Islands (see p. 46). The restored privileges of both the church and the aristocracy became the pivotal element in the so-called "Republic of the Seven United Islands." Several drafts of a constitution written between 1801 and 1803 defined Orthodoxy as the official religion and Greek as the official language.[19] Centrifugal forces within the Ottoman Empire gained momentum, so that the Ottoman Empire was politically weakened by the end of the eighteenth century, and the Janissaries, the Ottoman elite troop, were considered to be more of a burden than an asset. The political situation led to the reign of Maḥmūd II, who would become a great modernizer after the war against Russia, which was followed by the Serbian and the Greek uprising in 1820, triggered by a "hellenophile effusiveness" (Brockelmann 1943).[20]

16 About Austria's role in the Balkans and vis-à-vis the Ottoman Empire and Prussia, v. Helmut Kuzmics and Roland Axtmann, *Autorität, Staat und Nationalcharakter: Der Zivilisationsprozeß in Österreich und England, 1700–1900* (Opladen: Leske und Budrich, 2000), 111–2.
17 Clogg, "The ‚Dhidhaskalia Patriki'," op. cit., 91.
18 Cf. Kreutz, *Das Ende des levantinischen Zeitalters*, op. cit., 46–8. Vakalopoulos, *Griechische Geschichte*, op. cit., 93; Xydis, "Modern Greek Nationalism," op. cit., 232; Steven Runciman, *Das Patriarchat von Konstantinopel: vom Vorabend der türkischen Eroberung bis zum griechischen Unabhängigkeitskrieg* (Munich: Beck, 1970), 383. It was modeled after the Carboneria in Italy v. ibid. 232–3. The Carboneria secret society of which Giuseppe Mazzini (1805–72) was a member became a model for all political secret organizations in the nineteenth century, v. Rudolf Lill, *Geschichte Italiens vom 16. Jahrhundert bis zu den Anfängen des Faschismus* (Darmstadt: WBG, 1980), 100, 111; cf. Vakalopoulos, op. cit., 103. In Egypt, too, there were secret societies under Mazzini's influence; the most famous case in point was the Italian *Pensiero ed Azione*, v. Jacob M. Landau, "Prolegomena to a Study of Secret Societies in Modern Egypt," in idem, *Middle Eastern Themes: Papers in History and Politics*, London 1973 [= Reprint from *Middle Eastern Studies*, I (2), Jan. 1965): 7–56, here 9–10.
19 Konstantina Zanou, "Nostalgia, Self-Exile and the National Idea: The Case of Andrea Mustoxidi and the Early Nineteenth-Century Heptanesians of Italy," in *Nationalism in the Troubled Triangle: Cyprus, Greece and Turkey*, ed. Ayhan Aktar, Niyazi Kızılyürek and Umut Özkırımlı (London et al.: Palgrave Macmillan, 2010): 98–111, here 99.
20 Brockelmann, *Geschichte*, op. cit., 310–1.

Ioannis Kapodistrias, who would later become the first president of Greece, was a member of the Congress of Vienna (1814/15), which attempted to reorganize Europe after the end of Napoleon's rule.[21] Another Greek uprising against the Ottoman Empire broke out on the Peloponnese not long afterwards (on 25 March 1821), which was proclaimed by Archbishop Germanos, who would raise the Greek flag in Kalavrita (Peloponnese) during the first year of the uprising.[22] The uprising took a much more promising turn this time, which led to a stronger reaction from the Ottoman government. It hanged the Ecumenical Patriarch of Constantinople, Gregory V, above the gates of his palace, which have been closed ever since.[23] The struggle for independence then took a religious turn. As Maurer wrote, it became "a battle not only for political but at the same time for religious freedom."[24]

One of those freedom fighters was Ioannis Kolettis, a personal friend of Maurer's, who was in command of the so-called *palikaria*, i.e., a band of warriors. Kolettis had studied medicine in Italy and worked as Ali Pasha's private doctor in Ioannina.[25] The struggle for independence was accompanied by a massive upswing in educational and cultural renewal, with literary journals spreading all over the Eastern Mediterranean,[26] e.g., the *Literary Mercure* (ὁ λόγιος ἑρμῆς), edited by Anthimos Gazis.[27] It was also accompanied by the foundation of new educational institutions such as the Ionian Academy, which was established in 1807 under French occupation, and the Ionian University, funded by the philhellene Lord Guilford in 1823.[28]

The Greek insurrection between 1821 and 1830 was the first of its kind to end in an independent state, followed by Serbian, Romanian and Bulgarian independence in 1878, and Albanian independence in 1912. The new order of nation-states shaped according to ethnic boundaries was at least partly the result of the fact that revolutionary ideas were developed during the Enlightenment.[29]

---

21 Heinz Duchhardt, *Der Wiener Kongress: Die Neugestaltung Europas 1814/15* (Munich: Beck, 2013), 49–52, 88.
22 Von Maurer, *Das griechische Volk*, Erster Band [1st vol.], op. cit., 467–8.
23 Tamcke, op. cit., 20–1.
24 Von Maurer, *Das griechische Volk*, Erster Band [1st vol.], op. cit., 468.
25 Ibid., 439.
26 Ibid., 437.
27 Ibid., 432?; 438.
28 Ibid., 438.
29 Papoulia: "Με τη δημιουργία των κρατών αυτών έχουμε μια σαφή διαφοροποίηση από τις προηγούμενες εποχές. [...] Αυτό δεν σημαίνει ότι σε παλαιότερες εποχές δεν υπήρξαν στοιχεία εθνικής συνείδησης, κυρίως σε λαούς με έντονη πολιτιστική παράδοση – ορισμένοι λαοί, όπως οι έλληνες και οι Εβραίοι είχαν από πολύ παλιά συνείδηση της κοινής καταγωγής –

Von Maurer speaks of the "political rebirth" of the Greek people, reviving a consciousness of belonging that dates back to the antiquity.[30] This was chiefly the work of the Phanariots. The first grand dragoman of the Sublime Porte in the first half of the seventeenth century was a Phanariot, followed by the famous Alexandros Mavrokordatos, who would later design the first constitution in 1822.[31] This constitution, the so-called Constitution of Epidauros, was written in an archaic language, which meant that only a few educated Greeks of that time would understand it.[32]

It was not only for this reason that the constitution was a "stillborn child" (Maurer); it was also because it was meant only to be a means for Mavrokordatos to become Greece's ruler.[33] As Maurer states, another major problem under Ottoman occupation was the corruption of the judiciary, which rank-and-file clerics profited from in notarial recordings.[34] Ionos Dragoumis (1878–1920), a Greek nationalist at the turn of the century, would later argue that there had been a revolution after the revolution, when bishops and patriarchs lost their importance to prefects, ministers and the king, while communities lost their importance to municipalities. This did not go unnoticed by the Europeans, who realized that the Greeks were able to take their fate into their own hands.[35]

However, as Dragoumis also recalled, there was a swing back to older habits, when in Laconia (Peloponnese) the government allowed communities to rule

αλλά ότι η κρατική οργάνωση συμπίπτει ή πρέπει να συμπίπτει με τα εθνικά όρια. Αυτή είναι η λεγόμενη *αρχή των εθνοτήτων*, την οποία διακήρυξαν οι φορείς των επαναστατικών ιδεών κατά την *Εποχή των Φώτων* (Siècle des lumières) ή Διαφωτισμού. Papoulia, *Από την αυτοκρατορία στο εθνικό κράτος*, op. cit., 48. Cf. Emanuel Turczynski, "Gestaltwandel und Trägerschichten der Aufklärung in Ost- und Südosteuropa," in *Die Aufklärung in Ost- und Südosteuropa: Aufsätze, Vorträge, Diskussionen*, ed. Erna Lesky, Strahinja K. Kostić, Josef Matl and Georg von Rauch (Cologne and Vienna: Böhlau, 1972): 23–49, passim.

30 Von Maurer, *Das griechische Volk*, Erster Band [1st vol.], op. cit., 1.
31 Ibid., 92–3; idem., *Das griechische Volk*, Zweiter Band [2nd vol.], op. cit., 31–2,
32 Cf. Kreutz, *Das Ende des levantinischen Zeitalters*, op. cit., 49.
33 Von Maurer, *Das griechische Volk*, Zweiter Band [2nd vol.], op. cit., 31–2.
34 Von Maurer, *Das griechische Volk*, Erster Band [1st vol.], op. cit., 95.
35 Dragoumis: "'Όλα ἦταν πρίν συγυρισμένα, βαλμένα στή θέση τους, βυζαντινά ἀπομεινάρια, κατασταλαγμένη ζωή, ἀποτέλεσμα πιό παλιῶν πολιτισμῶν καί χρόνων. Ἔπειτα ἔξαφνα μέ τό 1821 ξελευτερώθηκε μιά πολιτεία ἑλληνική, εἶδε ὁ καθένας πως μποροῦν καί ἀλλάζουν τά πράματα, εἶδαν τοὺς Εὐρωπαίους, ἔφεραν καινούργια ροῦχα ἀπό τήν Εὐρώπη καί νέα συστήματα νά κυβερνοῦν τά κοινά πράγματα· ἀντί κοινότητες, ἔφεραν δήμους, ἀντί τό δεσπότη καί τόν πατριάρχη, εἶχαν τό νομάρχη, τόν ὑπουργό καί τό βασιλιά. Οἱ δάσκαλοι καί αὐτοί ἄλλαξαν θέση καί ἔγιναν λιγώτερο σπουδαῖοι. Κ' ἔγινε μιά πιό μεγάλη ἐπανάσταση ἔτσι, παρά ἡ ἐπανάσταση που εἶχε γίνει ἐναντίον τοῦ σουλτάνου." Ionos Dragoumis, Ὁ Ἑλληνισμός μου καί οἱ Ἕλληνες [1903–11] (Athens: Nea Thesis, 1991): chap. B, § 2, Athens 1904: 40–1.

themselves, but neglected to employ any police since they were deemed unnecessary and only appropriate for larger cities. Thus, the communities guarded their forests, shepherds, and orchards on their own, as was the case in Turkey, so that finally communities began to mushroom throughout the Greek state, which was the very opposite of the government's intention.[36] This can be explained by the Greek tendency in the late eighteenth century towards commerce, which, according to Maurer, made possible Korais and the Greek cultural revival,[37] a revival that was largely independent of the government. Dragoumis therefore argued that "the Greeks differ from other Romans because they are a commercial people."[38] This constitutes the background for the emerging public consciousness that a free nation would overcome illiteracy and promote education. Blame for the supposed backwardness was pinned on the church, which had long been loyal to the Ottoman masters.[39]

When Ibrāhīm, Muḥammad ʿAlī's son, led a campaign in Greece at the behest of the sultan, France, England and Russia, followed by Prussia, founded the Triple Alliance in 1827, which aimed to press the Ottomans to let go of the Greeks. After the Egyptians had conquered the Greeks' last line of defence, the citadel of Missolonghi, they faced a new military challenge. This challenge came from across the sea, where the united fleet of the Triple Alliance forbade Ibrāhīm to take any further action against the Greeks. When Ibrāhīm refused to accept this injunction, there was a clash between the navy of the Triple Alliance and that of the Ottomans, which had been sent there to support Ibrāhīm. But the Triple Alliance emerged victorious.[40] However, Russia began a campaign and finally forced the Ottoman government to accept the London Treaty granting Greece independence.[41]

The young intelligentsia had during the Greek war of independence become more radical and had rejected thinkers who were in line with Christianity.

---

36 Dragoumis: "[...] Τό κράτος θ' ἀφήση τίς κοινότητες νά διευθύνωνται [s. lenken] μόνες τους (τοπική αὐτοδιοίκηση). Ἕνας νομάρχης μόνο θά διορίζεται καί θά πληρώνεται ἀπό τό κράτος γιά νά ἐπιβλέπη. Καί οἱ χωροφύλακες μποροῦν νά λείψουν. Ἡ ἀστυνομία θά εἶναι γιά τίς μεγάλες πόλεις. Τά δάση τους, τίς βοσκές τους, τά περιβόλια τους θά τά φυλάγουν μόνες τους οἱ κοινότητες, ὅπως στήν Τουρκία. Ἡ κοινότης ξεφυτρώνει παντοῦ μέσα στό ἑλληνικό κράτος, που τήν εἶχε καταργήσει. [...]" Ibid., chap. B, § 16: 46.

37 Von Maurer, *Das griechische Volk*, Erster Band [1st vol.], op. cit., 439.

38 Dragoumis: "Οἱ Ἕλληνες διαφέρουν ἀπό τοῦς ἄλλους Λατίνους, γιατί εἶναι ἐμπορικός λαός." Dragoumis, Ὁ Ἑλληνισμός μου καί οἱ Ἕλληνες, op. cit., 122.

39 Apostolos Vacalopoulos, "Byzantinism and Hellenism," in *Balkan Studies*, Vol. 9 (1968): 101–26, here 31.

40 Brockelmann, *Geschichte*, op. cit., 311–2.

41 Ibid., 312.

Among the new generation of intellectuals who did not shy away from criticizing the church were men such as Theofilos Kairis (d. 1853), Theoklitos Farmakidis, Neofitos Vamvas, the Oekonomos brothers, and, the most influential of all, Adamantios Korais (d. 1833). Korais also had prominent adversaries, however, such as Neofitos Doukas, Katartzis, and Stefanos Kommitas.[42]

The Greek question that emerged soon after the war of independence was shaped by the principles of the Metternichean system and the ambitions of the Great Powers, i.e., England, France, Austria and Russia. It was against this background that the Greek constitutionalist Adamantios Korais sought to consolidate the Greek state by using the example of the US, and he therefore began to correspond with Thomas Jefferson. The advice that Jefferson gave was puzzling: instead of praising the American constitution, he emphasized the freedom of the press as "the best instrument for enlightening the mind of man."[43] However, there was no juridical terminology in modern Greek,[44] and, according to Maurer, the political system was a rather unsystematic imitation of the French codex of civil procedure.[45]

There had been a craving in Greece for more than fifty years for French education and French institutions, although, as, Maurer argued, the Germanic spirit and its institutions were more fitting.[46] Korais was not especially supportive of this idea, since he disliked the Germans for their "tendency to systems."[47] In any case, Greece seemed to be in a predicament. After "everything had been slowly turned over," Dragoumis noted, "nothing was left in place, which is how we find ourselves now." Unable to create a foundation for their national identity, the Greeks could not find a political "form that suits us since we are Greeks and not Chinese or Bulgarians." The old structures with an admixture of "Frankish

---

42 Hering, "Die Auseinandersetzungen über die neugriechische Schriftsprache," op. cit., 132–5; On Korais v. Maurer, *Das griechische Volk*, Erster Band [1st vol.], op. cit., 30, 433–4.
43 Stephen G. Chaconas, "The Jefferson–Korais Correspondence," in *The Journal of Modern History*, Vol. 14, No. 1 (March 1942): 64–70, here 64–5. Adamantios Koraes [Korais], de L. R. S. Chardon, and de P. W. Brunet, *Report on the Present State of Civilization in Greece*, in *Lettres inédites de Coray à Chardon de la Rochette* (Paris 1877): 451–90, after English translation in *Nationalism in Asia and Africa*, ed. Elie Kedourie (Trowbridge and London: Weidenfeld and Nicholson, 1970): 153–188.
44 Von Maurer, *Das griechische Volk*, Zweiter Band [2nd vol.], op. cit., 343.
45 Ibid., Erster Band [1st vol.], op. cit., 588.
46 Ibid., Zweiter Band [2nd vol.], op. cit., 345–7.
47 Noutsos, "Christian Wolff und die neugriechische Aufklärung," op. cit., 77.

innovations" were simply imposed on the Greeks, who were still searching for their own ideal.[48]

2.2 The End of Eastern Decline

The ideas of modernization were being imposed from the top on the other side of the Mediterranean shore, too. Egypt remained legally part of the Ottoman Empire until the beginning of the First World War.[49] It was Muḥammad ʿAlī, a former officer of Albanian extraction, who, at the behest of the Sublime Porte, clamped down on the unrest in Arabia between 1811 and 1819, conquered Sudan in 1820–1822, and fought the Greeks in 1822–1826, until modernizing the country's army became inevitable. Most of the troops were Egyptian in name, but commanded by foreign officers. The military schools that opened their gates in 1816 were open to all nationalities – except Egyptians.[50] Muḥammad ʿAlī founded a hospital and a medical school run by Bartholomew Clot (later known as "Clot Bey") for the medical treatment of his troops in 1826. Its staff consisted entirely of locals and was still managed by locals after the British occupied Egypt in 1882. The school had a difficult start. Since students who came from the Azhar were strangers to foreign languages and the teachers did not speak Arabic, interpreters were needed. But those employed to interpret knew nothing about medicine and it took a time for these problems to be solved.[51]

The programme of modernization would soon affect many different aspects of life. Muḥammad ʿAlī reduced feudalism (*iltizām*) and undertook land reform. However, the land was still concentrated in the hands of a relatively small number of owners.[52] Muḥammad ʿAlī's programme of industrialization had proved a failure, but he was more successful in agriculture. Egypt had a population by the middle of the nineteenth century of about two and a half million, about 80% of

48 "Καὶ ἀφοῦ ἀναποδογυρίζονταν σιγά σιγά ὅλα, τίποτε δέν ἔμεινε στή θέση του καί ἔτσι βρισκόμαστε τώρα, χωρίς νά μποροῦμε ναὑροῦμε τή νέα μορφή μας, τό καταστάλαγμά μας, τή μορφή που μᾶς πρέπει, ἀφοῦ εἴμαστε Ἕλληνες καὶ ὄχι Κινέζοι ἤ Βούλγαροι. [...] ἀπό τά παλιά πολλά εἶναι πεσμένα, ἄλλα μισοστέκονται, μά θά γκρεμιστοῦν καί αὐτά, ἄλλα ἔχουν ἀνακατωθῆ μέ φράγκικους νεωτερισμούς, ἄλλα εἶναι ὅλο φράγκικα, – καινούργια ἤ ἀποφόρια – καί δέν τά ἔχομε χωνέψει ἀκόμη. – Ποῦ ναὑρεθῆ τό ἑλληνικό ἰδανικό ἀνάμεσα σέ τόση ἀνακατωσούρα;" Dragoumis, Ὁ Ἑλληνισμός μου καί οἱ Ἕλληνες, op. cit., chap. B, § 2: 40–1.

49 J.N.D. Anderson, "Law Reform in Egypt, 1850–1950," in *Revolution in the Middle East and Other Case Studies*, ed. PJ Vatikiotis (London: Allen and Unwin, 1972): 146–172, here 149.

50 Antoine B. Zahlan, "The Impact of Technology Change," op. cit., 31–58, 33–4.

51 Ibid., 35.

52 Anderson, "Law Reform in Egypt," op. cit., 151.

whom were illiterate peasants.[53] The second half of the nineteenth century was marked by attempts to introduce elements of democracy.[54]Deist thinkers in Europe, such as John Spencer (1630–93), Ralph Cudworth (1617–89) and Johann Salomo Semler (1725–1791), had long been fascinated by the country of the Nile,[55] and European intellectuals were intrigued by Muḥammad ʿAlī's attempts to modernize Egypt and to push forward some kind of proto-globalization connecting East and West. This was especially true for the Saint-Simonists in France, who travelled to Egypt in 1833 to spread the idea of making peace through industrial interrelations aimed at overcoming the rift between East and West.[56]

One of those was the Frenchman Michel Chevalier (1806–1879), who praised Muḥammad ʿAlī for his conquest of Syria.[57] Chevalier described in his essay "Système de la Méditerranée" (1832) his vision of a Mediterranean world as a place of Oriental and Occidental entanglement, a place whose pacification should be the goal of politics.[58] Accordingly, the nations of the world would complete each other since each possesses unique characteristics.[59] Chevalier made Russia the bogeyman and viewed Russians as strangers to liberalism and as generally being alien to the idea of democracy. The Russians were a people, he concluded, who were too obedient, too imitative, and too passive.[60] Russia's antithesis was Egypt.

The Egyptian journal *al-Hilāl* praised Muḥammad ʿAlī in 1892 as being the last in a long line of Egyptians who were sources of wisdom, a line that stretched

53 Ibid., 146.

54 Ibid., 149–50.

55 Blumenberg, *Das Lachen der Thrakerin*, op. cit., 49–50; idem., *Arbeit am Mythos*, op. cit., 55. A similar view was taken by the German Enlightener August W. Schlözer, v. Mühlpfordt, "Hellas als Wegweiser zur Demokratie," op. cit., 251. Napoleon himself seems to have even looked for ways to present French deists as Muslims. Bonaparte's interest in Mohammad might be genuine, even if this was only for legitimacy purposes as Cole (2008) thinks, v. Cole, *Napoleon's Egypt*, op. cit., 129.

56 Alexander Schölch, "Der arabische Osten im neunzehnten Jahrhundert, 1800–1914," in *Geschichte der arabischen Welt*, ed. Ulrich Haarmann and Heinz Halm (Munich: Beck, 2004): 365–431, here 68; Israel Gershoni and James P. Jankowski, *Egypt, Islam, and the Arabs: The Search for Egyptian Nationhood, 1900–1930* (New York and Oxford: Oxford University Press, 1986), 114; Ismāʿīl Aḥmad Adham, "li-mādhā ana mulḥid?" [1937], in idem, *min maṣādir al-tārīkh al-islāmī wa-nuṣūṣ ukhrā* (Damaskus: Dar Petra, 2009): 157–6, here 159; Kreutz, *Das Ende des levantinischen Zeitalters*, op. cit., 96.

57 Michel Chevalier, *Système de la Méditerranée* (Paris: Manucius, 1832), 27; cf. Kreutz, *Das Ende des levantinischen Zeitalters*, op. cit., 136.

58 Chevalier, op. cit., 28; cf. Kreutz, op. cit., 64–5.

59 Chevalier, op. cit., 24; cf. Kreutz, op. cit., 187.

60 Chevalier, op. cit., 17; cf. Kreutz, op. cit., 69.

back to the earliest times and even to the period of Roman rule. It was the Arab statehood of Islam (*daulat al-Islām al-ʿarabīya*) that had brought Egypt from the rubble to the throne of glory. When it entered the realm of the Kurds and the Circassians, God made Egypt part of the Ottoman Empire. Napoleon's reign, the article continued, remained only a short episode before the Ottoman Empire rose again. Muḥammad ʿAlī's ascent to power ended the period of ignorance and began the period of reform (*tārīkh al-iṣlāḥ al-akhīr*) in Egypt, reform that affected agriculture, industry, commerce, politics, the judiciary, science, and other fields. Finally, the article concluded, Egypt was able to catch up with European countries in terms of administration, education and its reputation (*taraqqī shaʾnihā*).[61]

The article in *al-Hilāl* argued that the programme of modernization found vast support among the local population, since progress was seen as a matter of "prestige." It comprised reforms that had first been introduced by Europeans, who had founded factories and schools, reorganized the army, and achieved much else of historically outstanding character.[62] When the flag of liberty (*liwāʾ al-ḥurrīya*) was raised, the Egyptians began to enjoy personal freedom (*al-ḥurrīya ash-shakhṣīya*) – at least for a time.[63]

A French diplomatic source tells us that Muḥammad ʿAlī devoted himself to the idea of Egyptian civilization "as we understand the word 'civilization', in the sense of our modern ideas," and that the civilization of Egypt is attaching itself to this great progressive movement that is sweeping through both the Orient and the Occident. In the Christian states, the source continues, civilization used to be shaped by degree and as a natural result of the ideas that Christianity had introduced. Now, civilization is supposed to take root in the Muslim states, where nothing has been prepared for its arrival. The leaders of the Muslim world were therefore not on a par with the general level of progress found in other parts of the world, and had no choice but to introduce a gradual and inevitable programme of modernization in order to catch up with the modern nations of the time.[64]

When Muḥammad ʿAlī's grandson ʿAbbās rose to power in 1848, the medical school mentioned above became a bone of contention between the British and the French. Since the heads of the school at the time, the Frenchmen Durignéan and Perron, had been unsuccessful, ʿAbbās turned to Germany and employed as

61 N.N., "an-Nahḍa al-miṣrīya al-akhīra," in *al-Hilāl*, vol. I, 1/1, Sep. 1, 1892 (Cairo): 123–5, here 123.

62 Ibid., 123–4.

63 Ibid., 124.

64 MD Turquie, Tome 20 [S. 224] Alexandria, 3. Juli 1833, no. 23: De la civilization de l'Egypte, quoted after Kreutz, *Das Ende des levantinischen Zeitalters*, op. cit., 95.

the school's new head first Wilhelm Griesinger in 1850 and then Theodor Bilharz in 1856. English had replaced Arabic as the language of teaching by the end of the century and the staff now consisted entirely of British and German professors.[65] While the languages in Egypt had under Muḥammad ʿAlī been *de facto* Arabic and French, this changed under Saʿīd and Ismāʿīl, when Arabic was the only official language in the country.[66] Ismāʿīl, who acquired the title of a Khedive in 1867, pushed the intellectual Nahḍa even further.[67]

The Arabic Nahḍa was based on introducing new ideas into the public discourse,[68] while turning to the European Enlightenment, and especially to proponents of the French Revolution such as Rousseau, Montesquieu, and Descartes. The "East" achieved what Europe had already achieved.[69] As Barqāwī (1988) argues, the Western Enlightenment (*at-tanwīr al-gharbī*) was considered an abundant source from which Eastern intellectuals could gain so much inspiration for their own culture.[70] This process was supported by direct contacts with the West, so that elements of Western culture and thinking were steadily transmitted to the Arab world. However, the adoption of Western ideas during the Nahḍa did not occur in an uncritical or unconditional manner; rather, the ideas underwent a process of Arabization (*taʿrīb*) that promoted genuine and authentic progress while looking for the reasons behind the rise of the West.[71]

When the Ottoman sultan Maḥmūd decided to reorganize the army, he made use of instructors sent by Egypt's Muḥammad ʿAlī. The reluctant Janissaries in Constantinople were rounded up and murdered in a bloodbath in 1826. Their close allies, the dervish order of the Bektashi, were persecuted and forced to flee to the borders of the empire. This made reforms to the Ottoman state even more urgent, and the fact that Muḥammad ʿAlī had already begun this process of reform in Egypt placed additional pressure on Maḥmūd.[72]

---

65 Zahlan, "The Impact of Technology Change," op. cit., 35–6; v. also James Heyworth-Dunne, *An Introduction to the History of Education in Modern Egypt* (London: F. Cass, 1968), *passim*; cf. Kamāl Raḍwān, *Almān fī Miṣr* (Cairo: al-Maktaba al-Qawmiyya al-Thaqafiyya, 1979), 17–8, 76.

66 وفي عهد سعيد وإسماعيل أصبحت اللغة العربية اللغة الرسمية الوحيدة في البلاد Barqāwī, *muḥāwala*, op. cit., 23.

67 ولا سيما أن حركة التعليم قد نشطت نشاطاً بارزاً .فإن اسماعيل قد دفع بالنهضة الفكرية قدماً إلى الأمام Ibid., 23. Anderson, "Law Reform in Egypt," op. cit., 149.

68 According to Barqāwī also in peace and unity, v. Barqāwī, *muḥāwala*, op. cit., 115. This is in stark contrasts to Salvatore's contention that the Nahḍa will carry out a modern turn but cannot define its own aspirations, v. Salvatore, op. cit., 224–5.

69 Barqāwī, *muḥāwala*, op. cit., 33, cf. 36.

70 Ibid., 42.

71 Ibid., 44–5.

72 Brockelmann, *Geschichte*, op. cit., 311–3. On the Bektashis v. Kreutz, *Das Ende des levantinischen Zeitalters*, op. cit., 122, 196, 280.

The spirit of reform began when the Ottomans decided to modernize their army, a project that was partly sustained in its early stages by the French officer Baron de Tott.[73] the Tanẓīmāt reforms began in 1839 with the *Hatt-i Sherif* of Gülhane,[74] while Egypt under *Muḥammad ʿAlī had* its own reform policy.[75] The *Hatt-i Sherif* introduced a number of reforms, such as the equal treatment of all subjects irrespective of their religion and ethnicity, the guarantee of private property, and, as mentioned above, reform to the law so that non-Muslim witnesses were permitted in court. A commercial code based on the French model was implemented and torture was officially banned. The French revolutionary ideas of liberty, equality and fraternity were the leitmotifs in the background.[76]

According to Barqāwī (1988), the idea of Ottomanism is tied primarily to the state and not to the *umma*. It only becomes the Ottoman *umma* when it includes all ethnic groups and peoples.[77] When corruption led to political decline,[78] and to the crumbling of Ottomanism, the Empire gained new strength by implementing reforms.[79] The details of reform differed from place to place, dependent on local conditions and customs. This should be understood in conjunction with the major aim of the constitutional government, which was to achieve political stability.[80]

---

73 Bernard Lewis, "From Pilgrims to Tourists: A Survey of Middle Eastern Travel," in idem, *From Babel to Dragomans. Interpreting the Middle East* (London: Weidenfeld & Nicolson, 2004): 137–51, here 143.

74 Josef Matuz, *Das Osmanische Reich: Grundlinien seiner Geschichte* (Darmstadt: WBG, 1994), 225. Brockelmann, *Geschichte*, op. cit., 331, 324–5. Cook, *Ancient Religions*, op. cit. 177.

75 Anderson, "Law Reform in Egypt," op. cit., 147–8.

76 Matuz, *Das Osmanische Reich*, op. cit., 225, 230; Aron Rodrigue, "The Beginnings of Westernization and Community Reform Among Istanbul's Jewry, 1854–65," in *The Jews of the Ottoman Empire*, ed. Avigdor Levy (Princeton/New Jersey: The Darwin Press, 1994): 439–56, here 444; Hösch, *Geschichte der Balkanländer*, op. cit., 122–3. Muḥammad Saʿīd al-Usṭuwānī, *Mashāhid wa-aḥdāth dimashqīya fī muntaṣaf al-qarn al-tāsiʿ ʿashar: 1256/1840–1277/1861*, ed. Asʿad al-Usṭuwānī (Damascus: Dar al-Jumhuriyya, 1993), 162. On the influence of the French Revolution v. Axel Havemann, "Geschichte und Geschichtsschreibung im Libanon: Kamāl Ṣalībī und die nationale Identität," in *Gegenwart als Geschichte: Islamwissenschaftliche Studien: Fritz Steppat zum fünfundsechzigsten Geburtstag*, ed. Axel Havemann and Baber Johansen (Leiden et al.: Brill, 1988): 225–43, here 121.

77 Barqāwī, *muḥāwala*, op. cit. 77.

78 Sulaymān al-Bustānī, *ʿibra wa-dhikrā aw ad-daula al-ʿuthmānīya qabla al-dastūr wa-baʿdahu* (Beirut: Dar at-Taliʿa, [1904] 1978), 144. والدولة على ذلك الانحطاط لم تَعْدَم رجالا هذا شأنهم، عاش من عاش منهم في جهاد دائم ومات من مات حزينا اسيفا Ibid.

79 أما الدولة العثمانية، فبعد أن كانت هذه البلاد علة ضعف وفقر ومعرة لها، فستصبح ان شاء الله مورد قوة وغنى وفخر عظيم Ibid., 209.

80 وأما وسائل الاصلاح فهي، وأن اختلف بعضها في بقعة عما سواها، بالنظر الى طرق السقي وطبيعة البلاد وخلق السكان، فمرجعها كلها الى توطيد الامن والعناية بالفلاح، وقد اتضح جليا حتى الآن ان ذلك من أجل ما ترمي اليه حكومتنا الدستورية Ibid., 210.

This all is tied to the "Oriental Question," which is also relevant for how Europe understood itself at the time. The Russian conquest of the Crimea, which was a first step in breaking through to the Mediterranean shores, was made possible by the Treaty of Küçük Kaynarca (1774).[81] A kind of "nineteenth-century world war" (Diner 1996) took place between 1853 and 1856, one tied to the "Oriental Question:" Russia had lost the war against the Ottomans over the Crimea, and Austria's position in Europe appeared insecure in 1849. This constellation was the prelude to the later antagonism between Austria and Russia, and consequently to the First World War.[82] It was mirrored in the religious field, with the Orthodox and the Catholic clergy arguing over the rights to be present on the holy sites in Jerusalem, a question opened politically by Napoleon III and Nicolaus I.[83] At the center of this power play was Greece, which drew interest from four great powers: Russia, England, France and Austria.[84]

The Crimean War and the Treaty of Paris of 1856, which made Russia withdraw from some of its Eastern territories, meant for the Ottoman Empire growing closer to Europe and advancing its own infrastructure.[85] While Russia distanced itself from Europe after losing the Crimean War, the Ottoman Empire entered the circle of European power states and was integrated into the European system of public law as granted by the Treaty of 1856. The progress and productivity of Europe was said to be the outcome of the rule of law; hence, the interest of Middle Eastern peoples in constitutionalism.[86]

Russia pressed the Ottoman Empire to reform with the Treaty of San Stefano of 1878, and created an enlarged vassal state, including a large part of the Aegean and almost the whole of Macedonia. However, the European states opposed the expansion of Russia's influence. The Congress of Berlin of 1878 put an end to Russian aspirations, and made itself the guarantor of reforms in the Ottoman Empire. Macedonia remained under Ottoman sovereignty.[87]

The ideal of Europe grew popular in the eyes of many Muslims, but did not remained undisputed.[88] The ruling circles were more interested in modernization and more open to Western influence in Cairo than in Constantinople. But the

---

81 Kemal H. Karpat, *The Politicization of Islam: Reconstructing Identity, State, Faith, and Community in the Late Ottoman State* (Oxford: Oxford University Press, 2001), 72–3.
82 Dan Diner, "Zweierlei Osten," op. cit., 97–9; Karpat, *The Politicization of Islam*, op. cit., 75.
83 Dan Diner, "Zweierlei Osten," op. cit., 112.
84 Von Maurer, *Das griechische Volk*, Erster Band [1st vol.], op. cit., 25–8.
85 Karpat, *The Politicization of Islam*, op. cit., 73–4.
86 Dan Diner, "Zweierlei Osten," op. cit., 99, 100, 102–4.
87 Hosking, *Russland*, op. cit., 418.
88 Karpat, *The Politicization of Islam*, op. cit., 74.

idea of the *sharīʿa* (Islamic law) as the ideal law was not relinquished, although in Egypt there was a more pronounced differentiation between the two kinds of court, one secular and the other religious.[89] A less visible but even more important step had been Muḥammad ʿAlī's implementation of a new law, the *qānūn al-muntakhab*, in 1845 (*Code de Mehemet Ali*), which was based on the Mediterranean practice of finding a common law between members of different faiths or nationalities.[90]

The rift in the Egyptian juridical system deepened after 1875, when the *sharīʿa* was strictly confined to religious courts in terms of personal status law and the law of succession, while the code of law in regular courts was applied by lawyers trained in the European tradition. This marked a radical turn in Egypt's history.[91] The *École Khédievale de droit* was established in 1868, followed in 1875 by mixed courts (*sharīʿa* and *Code Napoléon*). Lawyers were systematically trained in Cairo from 1890; in Beirut, from 1913; and, in Damascus, from 1919.[92]

How could this new thinking be reconciled with Islam? Reformers such as Jamāladdīn al-Afghānī, Muḥammad ʿAbduh, and Rashīd Riḍā were all active in reforming Islam; in the case of the latter, this included affection for the Wahhābis.[93] Al-Afghānī was more a revolutionary than a reformer like ʿAbduh.[94] According to Barqāwī (1988), the Arabic Nahḍa, which was oriented towards the Arabic-Islamic heritage of the classical age and especially towards its rationalist heritage, witnessed a major religious reform with al-Afghānī."[95] This is a bold claim, since al-Afghānī called for a return to the foundations of religion in order to make the Muslim world great again, which is hardly comparable to Luther's intentions.[96]

---

89 Anderson, "Law Reform in Egypt," op. cit., 159–60.

90 The *qānūn* was an improved version of the *jamʿīya ḥaqqanīya* which had been implemented in the year before, i.e. in 1844. Ibid., 147–8.

91 Ibid., 155, 162; Jörn Thielmann, *Naṣr Ḥāmid Abū Zaid und die wiedererfundene ḥisba: Sharīʿa und Qānūn im heutigen Ägypten* (Würzburg: Ergon, 2003), 22, 70.

92 Tilman Nagel, *Zu den Grundlagen des islamischen Rechts* (Baden-Baden: Nomos, 2012), 8.

93 FO 141/489 [Kew London] Pan-Islamic Congress (Jerusalem), 23 January 1932, Zionist ambitions in Palestine: List of delegates who attended the Moslem Congress held at Jerusalem on the 16 December, 1931, quoted after Kreutz, *Das Ende des levantinischen Zeitalters*, op. cit., 245.

94 Barqāwī, *muḥāwala*, op. cit., 57.

95 Ibid., 31.

96 ʿAbarraḥmān ar-Rāfiʿī, *Jamāladdīn al-Afghānī: bāʿith nahḍat ash-sharq, 1838–1897* (Cairo: Dar al-Ma'arif, 1991), 76; Wolfgang Schluchter, *Die Entstehung des modernen Rationalismus: Eine Analyse von Max Webers Entwicklungsgeschichte des Okzidents* (Frankfurt/Main: Suhrkamp, 1998), 324–5.

Luther had of course a major impact on the German language, and language reform also played a major role within the Nahḍa.[97]

However, the crucial point is that Islamic jurisprudence was based very much on the idea that there is no free will; that Islam is very much focused on the other world; and that equality is a significantly higher value than freedom, which is a key concept in Enlightenment thinking.[98] What could not possibly be called Enlightenment thinking was when a nineteenth-century theologian such as Ḥusayn al-Jisr (1845–1909) tried to preserve the concept of predestination by watering it down and drawing on the teaching of *kasb* as being something that was between the extremes (*al-madhhab al-ʿadl al-mutawassiṭ*) and that enabled him to avoid fatalism (*iʿtiqād al-qadar wa-l-jabr*).[99] Al-Jisr would not in any case accept Western notions of freedom, which saw religion as being a legitimate target for criticism, which is the reason that he championed a concept of freedom that he called "balanced" (*muʿtadila*) in contrast to the supposed "deviant freedom" (*ḥurrīya tāʾiha*) of the West.[100]

For al-Jisr, the caliphate as a religious institution also contained a secular government. This attitude was typical of Hanafi theologians, who believed that the caliphate should be obeyed in every instance, and that opposing it deserved nothing less than the death penalty.[101] It may therefore come as a surprise that al-Jisr supported ʿAbdülḥamid's reform policy, since the constitution, which theoretically reduced the caliph's power, was at its core. Al-Jisr looked for ideas to help the Islamic world catch up with the West technologically and economically, and saw the main reasons for the West's state of progress in its combination of education and patriotism, which he equated with the idea of the public good.[102] But Sultan ʿAbdülḥamīd II, who was shortly afterwards forced to recognize the constitution of 1876, was toppled by parliament.[103]

---

97 Cf. Salāma Mūsā, *mā hiya an-nahḍa?* [1935] (Damascus: Dar al-Baath), 111.

98 Sharafī, *al-Islām wa-l-ḥurrīya*, op. cit., 125.

99 Ḥusayn al-Jisr, *Riyāḍ Ṭarābulus al-Shām min inshāʾ muḥarririhā*, 10 vols., ed. Muḥammad Kāmil al-Buḥairī (Tripoli, 1893–1901, vol. 8 [1900]), 59, 64–5, quoted after Johannes Ebert, *Religion und Reform in der arabischen Provinz: Ḥusayn Al-Gisr aṭ-Ṭarābulusī (1845–1909): Ein Islamischer Gelehrter zwischen Tradition und Reform* (Frankfurt/Main: Peter Lang, 1991), 119–20.

100 Al-Jisr, op. cit., vol. 3 (1895), 125, quoted after Ebert, *Religion und Reform*, op. cit, 116.

101 Al-Jisr, op. cit., vol. 1 (1893), 96–7, vol. 3 (1895), 125, quoted after Ebert, *Religion und Reform*, op. cit., 109.

102 Ebert, op. cit., 112–3.

103 Bustānī, *ʿibra wa-dhikrā*, op. cit., 153. Karpat, *The Politicization of Islam*, op. cit., 351; Brockelmann, *Geschichte*, op. cit., 345–6; Maher, "Umrisse einer neuen Kulturphilosophie in Ägypten," 313 Fn. 8. Kreutz, *Das Ende des levantinischen Zeitalters*, op. cit., 182.

The concept of freedom championed by al-Jisr had little to do with freedom. Freedom, as Isaiah Berlin reminded us, is simply what it is: a lack of coercion. It is "not equality or fairness or justice or culture, or human happiness or a quiet conscience ... it is a confusion of values to say that although my 'liberal', individual freedom may go by the board, some other kind of freedom – 'social' or 'economic' – is increased."[104] In contrast, al-Jisr endorsed slavery as an institution; in the context of Islam, slavery was practised and a slave might have been released in an act of clemency, but that did not question the institution as such.[105] We should remember that in the Islamic world slaves rose to the highest ranks, with the Mamlūks forming a dynasty that ruled Egypt for centuries.

At that time, Francis Fatḥallāh al-Marrāsh published his book *ghābat al-ḥaqq*, an allegorical dialogue on political freedom, in Syria in 1866.[106] Now the past looked even more gloomy. Faraḥ Anṭūn (1874–1922) published an article in the journal *al-Baṣīr* in 1898 with the title "The Decline of the Orient" (*inḥiṭāṭ ash-sharq*), in which he depicted the Orient in a purely negative way. All the peoples of the Orient, he wrote, were destined to decline due to the Asian climate, which was constant and pleasant, and thus helped visionaries and prophets. In contrast, the Occident's climate was more extreme and challenging, which is the reason that it fostered more initiative among its inhabitants.[107] In as late as the 1950s, the Egyptian author Salāma Mūsā (1887–1958) wrote that "the Arab *literati* need the Western Enlightenment for their Oriental minds,"[108] and that "the Arab mind needs a change in our time."[109]

The individual to take the reforms to a higher level was Midḥat Pasha, who became provincial governor of Baghdad in 1869 and paved the way for education, be it by establishing schools or founding journals, lifting censorship or rewarding authors. The Nahḍa of knowledge and literature found its peak at that

104 Isaiah Berlin, "Two Concepts of Liberty," in idem, *Four Essays on Liberty* (London, Oxford and New York: Oxford University Press, 1969): 118–72, here 125–6.
105 Ḥusayn al-Jisr, "al-risāla al-ḥamīdīya fī ḥaqīqat ad-diyāna al-islāmīya wa-ḥaqīqat ash-sharī'a al-muḥammadīya," 2nd 1933/4: 328–30, quoted after Ebert, *Religion und Reform*, op. cit., 104–5. On slavery (*riqq*) v. Sharafī, *al-Islām wa-l-ḥurrīya*, op. cit., 125.
106 Hisham Sharabi, *al-muthaqqafūn al-'arab wa-l-gharb: 'aṣr an-nahḍa 1875–1914* (Beirut: Dar an-Nahar, 1991), 146; Bernard Lewis, "The Idea of Political Freedom in Modern Islamic Political Thought," in idem, *Islam in History: Ideas, Men and Events in the Middle East* (London: Alcove Press, 1973): 267–281, here 273–4.
107 Christoph Herzog, "Zum Niedergangsdiskurs im Osmanischen Reich," in *Mythen, Geschichte(n), Identitäten: Der Kampf um die Vergangenheit*, ed. Stephan Conermann (Hamburg: EB-Verlag, 1999): 69–90, here 84–5.
108 والقراء العرب يحتاجون الى التنوير الغربي لعقولهم الشرقية Mūsā, *mā hiya an-nahḍa?*, op. cit., 156.
109 إن الذهن العربي في حاجة إلى أن يتغير في عصرنا Ibid., 157.

time.[110] Pasha also set an example of interreligious cooperation when in 1879 he supported appointing Mercado Elcalay to the post of chief rabbi. Elcalay had been the chief rabbi of Serbia when Pasha himself had been governor (*wālī*) of that province. Their careers declined later on, with both leaving the province during the early 1880s.[111]

The Lebanese-Ottoman politician and author Sulaymān al-Bustānī gives us an account of Pasha's achievements in Baghdad, "which became an outstanding place in the whole province and even more impressive than the capital. And that is because he set up the first printing press there; published a paper that he called *az-Zawrā'* (an old name for Baghdad); reformed the Oman maritime administration, which ran steamers between Baghdad and Basra, and from there to Yemen and the Hijaz; created the great iron plant; established the Trades Office; spread the spirit of helping one another in building companies; set up a company with the people of Baghdad; and established the tramway between Baghdad and al-Kadhim, the first tramway company in the Ottoman Empire as far as we know."[112] Bustānī praised the amount of liberty and loyalty that Midḥat gave to his governorate, and his efforts to "sow the seeds of freedom so that people act in favor of liberty and speak out for liberty whatever the situation if they are on the right side."[113]

As Barqāwī pointed out, there is a link between political-social reform and religious reform. They are interconnected, just as the downfall of the state and the downfall of religion are interconnected.[114] While most Islamic religious reformers agree that the constitutional system is the best system of rule, the Islam-

110 ونشط اصحاب الاقلام فأنشأوا .فمهد سبل التعليم، وفي زمنه انشئت المدارس الكثيرة، وظهرت في سوريا اول المجلات العربية صحف الاخبار، ووسع لهم نطاق الحرية في التحرير، وكافأ المؤلفين بمال بعضه من عنده وبعضه مما كان يرد من الاستانة بناء على اشارة منه .وكانت في زمنه نهضة للعلم والادب لا يزال كهول السوريين يتغنون بها Bustānī, *'ibra wa-dhikrā*, op. cit., 135–6. Brockelmann, *Geschichte*, op. cit., 337.

111 Christina Weber, *Die jüdische Gemeinde im Damaskus des 19. Jahrhunderts: Städtische Sozialgeschichte und osmanische Gerichtsbarkeit im Spiegel islamischer und jüdischer Quellen* (Berlin: Klaus Schwarz, 2011), 78; v. also Y. Harel, "Midhat Pasha and the Jewish Community of Damascus: Two New Documents," in *Turcica* 28 (1996): 339–46, here 340.

112 وان له فوق ذلك من الآثار في تلك الولاية القاصية في اطراف البلاد ما جعل بغداد تفاخر سائر الولايات حتى ما جاور منها عاصمة الملك، فهو الذي انشأ اول مطبعة في بغداد وأصدر فيها جريدة دعاها "الزوراء"، وهو الذي اصلح ادارة عمان البحرية التي اخذت تسير البواخر بين بغداد والبصرة ومنها الى اليمن والحجاز وهو الذي انشأ معمل الحديد الكبير والحقه بتلك الادارة، وهو ايضا انشأ مكتب الصنائع، وبث في البلاد روح التضافر على تأليف الشركات، فألف شركة من أهالي بغداد فانشأت طريق الترامواي بين بغداد والكاظم وهي اول شركة ترامواي في الولايات العثمانية على ما نعلم Bustānī, *'ibra wa-dhikrā*, op. cit., 132.

113 واطلق من الحرية لمأموريه بقدر ما القي عليهم من التبعة واوجب عليهم عدم المحاذرة من شئ اذا كانوا على ثقة من عملهم حتى لقد كان يوبخ المأمور الذي يأنس منه تزلفا اليه بقول او بفعل، وكان لا يدخر وسعا في القاء بذور الحرية ليألف الناس العمل بها والنطق بها مهما كانت الحال اذا كانوا في جانب الحق Ibid., 133.

114 Barqāwī, *muḥāwala*, op. cit., 50.

ic world still struggles with the concept of secularism. An exception is ʿAlī ʿAbdarrāziq's 1925 book on *al-islām wa-uṣūl al-ḥukm*,[115] which argues that Islam is a religion of inwardness and has no governmental system – a claim that would cost the author his life.[116] In contrast, proponents of pan-Islamism such as al-Afghānī (1838–1897) never presented a clear Islamic theory of reform.[117] Especially the apologetic modernism that championed harmony between the rule of law and the tenets of Islam failed in practice.[118]

Voices of more radical reform began to be heard at around the turn of the century, most notably those of reformers such as Muḥammad Shākir, Qāsim Amīn and Muḥammad ʿAbduh.[119] As already mentioned, Qāsim Amīn (1863–1908) blamed distortions to the Islamic religion for the decline of the Islamic world. Religion had evolved over time partly due to interpretation and transformation by scholars and jurists. For Amīn, the main reason for decline was Muslim ignorance of their own religion. He pointed to Quran 7,50–1, which promised hell for infidels, since they "took their religion for a sport and pastime," and "whom the life of the world beguiled" (translated by Mishari-Rashid).[120]

After the Islamic sciences moved to Europe, Amīn argued, the lamp of science went out in the East and scholars of Islam displayed no intellectual interest in matters beyond *kalām*, grammar and some other skills useful to Quranic studies. They were therefore no longer able to understand true Islam, but felt that their own weaknesses prevented them from advancing their minds with regard to religion.[121] Amīn did not "see anything more surprising than the current situation: Do we live for the past or for the future? Do we want to advance or to lag behind? We see the world in permanent revolution and its matters in constant change while we watch these mutations with a glazed eye, startled thought and a dazed soul that we are not aware what we are doing. Then we become lost to the past, seeking for a safe place in it and demanding help from it because we always seclude ourselves disappointedly."[122]

---

115 Ibid., 56.
116 Bassam Tibi, *Der Islam und das Problem der kulturellen Bewältigung sozialen Wandels* (Frankfurt/Main: Suhrkamp, 1985), 95.
117 Barqāwī, *muḥāwala*, op. cit., 56.
118 Lazarus-Yafeh, "Die islamische Reaktion," op. cit., 219.
119 Anderson, "Law Reform in Egypt," op. cit., 163.
120 اتخذوا دينهم لهواً ولعباً وغرتهم الحياة الدنيا Qāsim Amīn, *taḥrīr al-marʾa* ([Cairo 1899] Damascus: Dar al-Baath, s.a.), 115–6.
121 Ibid., 118.
122 لا أرى أعجب من حالنا: هل نعيش للماضي أو للمستقبل؟ هل نريد أن نتقدم أو نريد أن نتأخر؟ نرى العالم في تقلب مستمر وشؤونه في تغير دائم ونحن ننظر إلى ما يقع فيه من تبدل الأحوال بعين شاخصة وفكرة حائرة ونفس ذاهلة لا ندري ماذا نصنع، ثم ننهزم إلى الماضي نلتمس فيه مخلصاً ونطلب منه عوناً فنرتد دائماً خائبين Ibid., 182.

A similar discourse of decline took place in the Greek context. In order to understand this discourse, we should take a look at history. The relationship between the Eastern Roman Empire and the church was that of a "symphony," a concept introduced by Emperor Justinian.[123] This relationship was much closer than in the Latin West, where the church enjoyed greater independence, which caused more tension with the political sphere.[124] In contrast, Eastern Roman emperors such as Justinian regarded themselves as educators of the people, protectors of the polity, and bearers of virtue (*aretē*). Wakefulness was the manifestation of the good emperor, who took care of his people by day and by night.[125] At the same time, the emperor was not above the law and was not to interfere in the process of jurisdiction.[126] The institution of the emperor "was sent to the people by God as a living law."[127]The declaration of independence of the Greek Church on 28 July 1833 was designed to return it to the position that it had had before the Ottoman conquest. In terms of organization, though, it was modelled after the Russian Church, which had its own, independent synod with a president as its head instead of a patriarch.[128] The Church of Hellas jettisoned the Ecumenical Patriarchate when the latter was said to be too hesitant in his support of the Greek struggle for independence. This step was taken by Georg von Maurer, who ruled Greece when Otto of Bavaria was king.[129] Greece became one of the

123 Manfred Clauss, "Die συμφών von Kirche und Staat zur Zeit Justinians," in *Klassisches Altertum, Spätantike und frühes Christentum: Adolf Lippold zum 65. Geburtstag gewidmet*, ed. Karlheinz Dietz, Dieter Hennig and Hans Kaletsch (Würzburg: Der christliche Osten, 1993), 579–93, here 580.

124 Vasilios N. Makrides, "Orthodoxes Ost- und Südosteuropa: Ausnahmefall oder Besonderheit?" in *Die Vielfalt Europas: Identitäten und Räume: Beiträge einer internationalen Konferenz, Leipzig, 6.bis 9. Juni 2007*, ed. Winfried Eberhard (Leipzig: Leipziger Universitäts-Verlag, 2009): 203–18, here 212.

125 ὥσπερ ἀεὶ ἡ τῆς ἀρετῆς δύναμις ἐν τοῖς ἐαντίοις διαφαίνεται, οὕτω καὶ ἡ βασιλικὴ πρόνοιά τε καὶ διοίκησις ἐν ταῖς τῶν ὑπηκόων ἐνστάσεσι φανεροῦνται. (...) εἰ δὲ ἤ τὸ τῶν ἀνθρωπίνων πραγμάτων εὐμετάβλητον ἢ ἡ τοῦ θείου νεύματος κίνησις τοῖς ἀνθρωπίνοις ἐνσκήπτει κακοῖς, ἡ ἐπαγομένη ἄνωθεν μετὰ φιλανθρωπίας παιδεία τῆς βασιλικῆς προνοίας τε καὶ φιαλνθρωπίας ὑπόθεσις γίνεται. Edict 7 (Nov., Corpus Iuris III), quoted after Wilhelm Schubart, *Justinian und Theodora* (Munich: F. Bruckmann, 1943), 271 note 23, transl. op. cit., 39. About the motive of wakefulness v. ibid.

126 καὶ κἂν εἰ συμβαίη κέλευσιν ἡμετέραν ἐν μέσῳ κἂν εἰ θεῖον τύπον, κἂν εἰ πραγματικὸς εἴη, φοιτῆσαι λέγοντα τοιώσδε χρῆναι τὴν δίκην τεμεῖν, ἀκολουθείτο τῷ νόμῳ· ἡμεῖς γὰρ ἐκεῖνο βουλόμεθα κρατεῖν, ὅσπερ οἱ ἡμέτεροι βούλονται νόμοι. Nov. 82,13, quoted after Schubart, op. cit., 272 note 32, transl. op. cit., 41.

127 Schubart, op. cit., 272 note 32, transl. op. cit., 41.

128 Von Maurer, *Das griechische Volk*, Zweiter Band [2nd vol.], op. cit., 161.

129 Ibid., Erster Band [1st vol.], op. cit., 441.

first electoral democracies in Europe, even before Britain, under the reign of the Bavarian king. But, as Fukuyama (2014) points out, Greece, unlike Britain, lacked a bourgeois class of landowners, which made the country strongly influenced by patron-client relationships.[130]

In fact, as Maurer had long argued, the church grew rich on real estate. He estimated that about a quarter of land property belonged to churches and monasteries. At the same time, he wrote, most of the clergy were illiterate. Moreover, the clergy exerted a great influence on worldly matters, so that a citizen often consulted a cleric when he faced losing a business deal.[131] In Ottoman Greece, the state had almost the entire land under its control, so that, with the exception of a few cities and provinces, there was no private landownership.[132] Farmers cultivated the land owned by the Ottoman government, the *çıftlık* (*tziftliki*), and had to pay the tithe. The Greek population fell into two camps: farmers and soldiers.[133]

There was less governmental control at the border areas of the Ottoman Empire, and this is where we find the Greek *kefalochoria*, i.e., villages with free landed property belonging to the government, and not the community. There also used to be Greek *kefalochoria* in Albania, but they disappeared mostly under Ali Pasha.[134] Rumelia and the islands also enjoyed greater freedom,[135] and it is little wonder that Rumelia became the fatherland of the *palikaria*, headed by the chief captains, the so-called *armatoloi*.[136] The *palikaria*, knights without conscription, looked down their noses on unarmed people and their professions, which is the reason that, despite the shortage of labor, so many people did not work.[137]

In the past, real estate had been judged according to Islamic law, which was still important even after the liberation. Here, legislation came into play.[138] Both common sense and fairness were in 1830 made the basis of law, which was soon criticized for introducing arbitrariness into jurisdiction.[139] Loukas Argyropoulos began translating the Islamic codes of law from Ottoman Turkish into

---

130 Fukuyama, *Political Order and Political Decay*, op. cit., 101–2.
131 Von Maurer, *Das griechische Volk*, Zweiter Band [2nd vol.], op. cit., 26–7.
132 Ibid., Erster Band [1st vol.], op. cit., 44–5.
133 Ibid., 45–6.
134 Ibid., 45.
135 Ibid., Zweiter Band [2nd vol.], op. cit., 30.
136 Ibid., Erster Band [1st vol.], op. cit., 45.
137 Ibid., Zweiter Band [2nd vol.], op. cit., 28–9.
138 Ibid., Erster Band [1st vol.], op. cit., 553 [§ 221].
139 Ibid., 553.

Greek, but failed to complete his work due to the lack of appropriate dictionaries.[140] Greece had been without courts since October 1833, while the government spoke of "a return to the state of law."[141] At the same time, the French *code de commerce* was still in use. There was a translation of the *code* into Greek, but it was of poor quality, and a new translation was ordered together with a revised law. It was intended to be published in 1834.[142]

According to Maurer, there was a general complaint that the entire jurisdiction was a "slave to politics and public authority," and that "freedom and honour, the lives and property of people, had fallen prey to those who had arbitrary power over day-to-day decisions"[143] in the early post-revolutionary era.[144] This is a phenomenon that arose from the Ottoman heritage. The British consular officer James Henry Skene described in the middle of the nineteenth century how difficult it was to implement the rule of law in the Arab provinces of the Ottoman Empire.[145] The juridical system in Aleppo was completely defunct. Judges who were sent by the government in Constantinople used their term of office to do nothing else but enrich themselves.[146] Trials took place without testimony, or witnesses were bribed like the judges. Witnesses willing to take bribes were easy to find since they queued in front of the court. Bribes were also the main source of income for judges and prosecutors, who were not paid by the government.[147]

According to Maurer, "judges left the courts because they were poorly paid, so that finally the courts themselves had to close for lack of good laws and judges, in the hope that the king and his regency would soon arrive."[148] Maurer pleaded for the creation of free institutions to establish a grounded monarchical state in Greece, which would later be realized in the field of local administration.[149] State and church were separated when the church's *de facto* independence reached the status of a fully-fledged autocephaly in 1855. This was promoted by Theoklitos Farmakidis, a theologian trained in the German town of

140 Ibid., Zweiter Band [2nd vol.], op. cit., 334.
141 Ibid., Erster Band [1st vol.], op. cit., 594.
142 Ibid., Zweiter Band [2nd vol.], op. cit., 335.
143 Ibid., Erster Band [1st vol.], op. cit., 589–90.
144 Ibid., Zweiter Band [2nd vol.], op. cit., 338.
145 FO 861/7 [Kew London] Aleppo 1859–1880, Entry book: Report by J. H. Skene, from: Michael Kreutz, "Unrest at the Gates of Aleppo: British Perspectives on Bedouins in Northern Syria, 1848–1913," in *Journal of Levantine Studies* Vol. 2/2 (2012): 105–29, here 112–3; idem, *Das Ende des levantinischen Zeitalters*, op. cit., 87–8.
146 Kreutz, *Das Ende des levantinischen Zeitalters*, op. cit., 87–8.
147 Ibid., 87–8.
148 Von Maurer, *Das griechische Volk*, Erster Band [1st vol.], op. cit., 592.
149 Ibid., Zweiter Band [2nd vol.], op. cit., 88.

Göttingen.[150] And, yet, as we will soon see, there remained many ties between church, state, and society.

2.3 A Feminist Revolution

In a general atmosphere of reform and modernization urban societies underwent several changes. One of the most important changes undergone by urban societies in the general atmosphere of reform and modernization was the redefined role of women. While women were often portrayed in the Greek antiquity as "evil" (κακόν),[151] a change began during the eighteenth century, when the Venetian itinerate apostle Cosmas of Aetolia (1714–1779), a student of Voulgaris' and a fervent supporter of social equality, criticized the exploitation of women by wealthy merchants and called for the rights of women.[152] The Greek author Ionos Dragoumis wrote half a century later: "[t]he Greeks before 1821 had a life with an ideal. They led an Oriental life, with the purpose of being freed from the Turks and taking the city (i.e., Constantinople) back. After 1821, they considered rejecting various idols and the Oriental life." And what was that "Oriental life"? It was "women shut up in their homes," the "separation of women and men," and, finally, the ubiquity of religion ("the many churches, many icons, crosses and many prostrations, many fasts").[153]

There is an Islamic tradition of regarding women as *fitna*, meaning "temptation," but also "sedition" and "civil strife." In two passages, the Quran associates *fitna* with Satan, while the Sunnah connects Satan with women: *an-nisāʾ ḥabāʾil ash-shayṭān bi-hinna yuṣīdu r-rijāl* ("Women are the ropes of Satan with which he chases men"). The Prophet is also supposed to have said: *mā taraktu baʿdī fitnatan aḍarr ʿalā r-rajul min an-nisāʾ* ("I have left no temptation more harmful to

---

150 *Theologische Realenzyklopädie*, ed. Gerhard Müller, Pt. I, vol 3 (17) (Berlin: De Gruyter, 1993), 214.

151 Gotthard Strohmaier, *Ethical Sentences and Anecdotes of Greek Philosophes in Arabic Tradition*, in idem, *Von Demokrit bis Dante* (Hildesheim et al.: Olms, 1996), 44–52, here 46.

152 Podskalsky, *Griechische Theologie*, op. cit., 343–4.

153 Οἱ Ἕλληνες πρίν ἀπό τό 1821 εἶχαν μιά ζωή, μ' ἕνα ἰδανικό. Εἶχαν μιά ζωή ἀνατολίτικη, μέ τό σκοπό νά ἐλευθερωθοῦν ἀπό τούς Τούρκους καί νά τούς ξαναπάρουν τήν Πόλη. Ὕστερα ἀπό τό 1821 συλλογίστηκαν νά ρίξουν κάτω διάφορα εἴδωλα, που τά εἶχαν πρίν γιά ἀπείραχτα καί ἀλάθητα, π.χ. τήν ἀνατολίτικη ζωή μέ τίς ἰδέες πού ξανοίγουν ἀπό τήν ζωήν ἐκείνη, – οἱ γυναῖκες κλεισμένες στά σπίτια – χωρισμός τῶν γυκαικῶν ἀπό τούς ἄνδρες – ὄχι περίπατος – πολλή ἐκκλησία, πολλά είκονίσματα, πολλούς σταυρούς κάι μετάνοιες πολλές νηστεῖες. Dragoumis, Ὁ Ἑλληνισμός μου καί οἱ Ἕλληνες, op. cit., chap. B, § 2: 40–1.

men than women"). Scholars of *fiqh* have argued on the basis of these two sayings that women should stay at home.[154]

According to Lewis (2010), women "have played a much more important role in Muslim history than is usually conceded by historians. But they were until very recently precluded from contributing to the development of their society in the way that a succession of remarkable women have contributed to the flowering of the West."[155] The claim made by Cole (2008) that in Egypt the seclusion of women has been practised chiefly by the ruling class is therefore unlikely. Since seclusion required an additional number of servants, the absence of women in the house appeared as a class distinction. Not until French soldiers went on patrol in the streets did Egyptians begin to lock their wives in their homes and take to the streets in protest against the French.[156] Cole believes that being locked away made Muslim women in Egypt even more attractive to the French.[157] This is a rather dubious assumption when you cannot feel attracted to someone who is out of sight.

According to Hartmann (1851–1918), the background to the veil was entirely different. While men adopted new styles of fashion *alla franca* in the first decades of the nineteenth century,[158] women opted for more conservative clothing in the second half, when the prevailing fashion among women in Istanbul's Pera neighborhood was the so-called *ferece-yaşmak*, a mostly light-colored loose overcoat with a fine tulle which concealed the lower part of the face from the nose. Since this type of clothing tended to slip and tear, women would soon prefer the *arşaf-mendil*, i.e., a dark, belted coat, combined with an equally dark face veil imported from Syria. There were separate distribution channels for the two sets of clothing at that time. Since the sellers of *ferece-yaşmak*s did not sell *arşaf-mendil*s, they tried to secure their business by calling for a dress code in their favor. But female customers had long since made their choice. Hartmann was convinced that the veil issue would resolve itself once women rebelled against their seclusion.[159]

---

154 Assem Hefny, *Herrschaft und Islam: Religiös-politische Termini im Verständnis ägyptischer Autoren* (Frankfurt/Main: Peter Lang, 2014), 74–5.
155 Lewis, *Faith and Power*, op. cit., 67.
156 Cole, *Napoleon's Egypt*, op. cit., 134.
157 Ibid., 138.
158 Oliver Jens Schmitt, *Levantiner: Lebenswelten und Identitäten einer ethnokonfessionellen Gruppe im osmanischen Reich im ‚langen 19. Jahrhundert'* (München: R. Oldenbourg, 2005), 325; Kreutz, *Das Ende des levantinischen Zeitalters*, op. cit., 269.
159 Martin Hartmann, *Der Islamische Orient: Berichte und Forschungen*, Vol. III: *Unpolitische Briefe aus der Türkei* (Leipzig: Rudolf Haupt, 1910), 143; Kreutz, *Das Ende des levantinischen Zeitalters*, op. cit., 270.

Is the veil Islamic? Qāsim Amīn (1863–1908), author of a well-known 1899 treatise on the *Liberation of the Woman* (*taḥrīr al-mar'a*), saw no religious influence at stake. The Westerner, who loved to ascribe everything good to his religion, he wrote, believed that the Western woman's rise was due to the fact that her religion supported her to gain freedom. But that was not true. In reality, the Christian religion did not develop a system to ensure the freedom of women, and did not support their rights through private or public regulations.[160] According to Amīn, the "Islamic law sooner than other laws established the equality of the woman and announced her freedom and independence at a time when her position was at the lowest among all nations, and granted her full human rights." The woman therefore had the right to buy and sell, and to donate and inherit, without having to seek permission from her father or husband.[161]

Amīn moved from this observation to the realm of politics. Since unlimited rule led to abuse in the political system, it needed to be checked, discussed and monitored. Yet, the Islamic nations had for centuries been under the rule of an absolute tyranny, and their rulers had mostly used religion for their own ends. Nations ruled by despotism suffered from a decline in manners, and hence from an atmosphere in which people were deterred only by fear and acted only by way of force. It was for this reason that men suppressed women's rights and disdained them, and that women – whatever their position in a family – lived in despair. The man, Amīn concluded, enjoyed all the freedom, while the woman suffered under slavery; he had knowledge, while she had ignorance.[162]

Although Amīn wanted to improve women's status, his line of argumentation is somewhat odd. For him, women should learn everything that men learn because they cannot maintain their houses without acquiring some degree of rational and behavioural skill, or at least basic education,[163] by which he

160 والغربي الذي يحب أن يَنسِب كل شئ حَسَن الى دينه يعتقد أن المرأة الغربية ترقت لأن دينها المسيحي ساعدها على نَيْل حريتها. ولكن هذا الاعتقاد باطل. فإن الدين المسيحي لم يتعرض لوضع نظام يكْفُل حرية المرأة ولم يبين حقوقها بأحكام خاصة أو عامة. Amīn, *taḥrīr al-mar'a*, op. cit., 20.

161 سبَق الشرع الإسلامي كلَّ شريعةٍ سِواهُ في تقرير مساواة المرأة للرجل فأعلن حريتها واستقلالها يوم كانت في حضيض الانحطاط عند جميع الامم وخَولها كل حقوق الإنسان Ibid., 20–1.

162 من المجرب أن السلطة غير المحدودة تُغري بسوء الاستعمال إذا لم تجد حداً تقف امامه ورأياً يناقشها وهيئة تراقبها. ولهذا مضت القرون على الأمم الإسلامية وهي تحت حكم الاستبداد المطلق، وأساء حكامها في التصرف، وبالغوا في اتباع أهوائهم، واللعب بشؤون الرعية. بل لعبوا بالدين نفسه في أغلب الأزمنة. ولا يستثنى إلا عدد قليل لا يكاد يذكر بالنسبة إلى غالبهم. إذا غلب الاستبداد على أمة لم يقف أثره في الأنفس عندما هو في نفس الحاكم الأعلى ولكنه يتصل منه بمن حوله ومنهم إلى من دونهم وينفُث روحه في كل قوي بالنسبة لكل ضعيف متى مكنته القوة من التحكم فيه. وكان من أثر هذه الحكومات الاستبدادية أن الرجل في قوته أخذ يحتقر المرأة في ضعفها ... وقد يكون من أسباب ذلك أن أول أثر يظهر في الأمة المحكومة بالاستبداد هو فساد الأخلاق ... فمن طبيعة هذه الحالة أن الإنسان لا يحترم إلا القوة ولا يُرْدَع إلا بلاخوف Ibid., 22–4.

163 Ibid., 28; cf. 55 where he confirms that complete equality is not necessary but at least elementary education.

meant reading and writing.[164] But this is only the starting-point. Amīn believed that nothing prevented the Egyptian woman, like the Western woman, from being involved in science, literature, the arts, commerce or handcraft, and that it was only her remaining in ignorance and neglecting her education that prevented her doing so.[165] Amīn viewed the education of women as pivotal in raising children, since the education that children received came from the women around them.[166]

To defend women's education, Amīn refered to the West, where it had become a greater success than the education of men. Westerners were proud that women participated in more aspects of life than ever before. Amīn quoted Ernest Renan (1823–1892) as having said that the most beautiful things that he had written were inspired by his sister.[167] According to Amīn, the assumption of some Muslims that the *sharia* commands women to wear the veil is rather dubious. If it were a divine command, then any discussion of the issue would be prohibited. The veil was not prescribed by the *sharia*; rather, it was a custom that Muslims had adopted as a result of cultural contact with other peoples, a custom that they had transformed into a religious duty. Like other negative customs, the veil had become associated with religion without having any base in it.[168]

Amīn championed lifting the veil not because he wanted to emulate the West, but because he took the *sharia* seriously.[169] The veil, he argued, was not specific to a certain culture or religion, but had been common among different cultures of the world, both Muslim and non-Muslim, before vanishing in the course of history.[170] While he regarded marriage sanctioned by the divine as the only legitimate way for a man and a woman to live together,[171] he rejected as baseless the claim that the veil was an issue of manners (*ādāb*). Why should there be a connection between good behaviour (*adab*) and veiling the face, he asked, since there was no reason to believe that good behaviour was not a quality both of men *and* women. The fear of seduction (*fitna*), which is so widespread in Islamic cultural and religious history, is also an invalid argument because it is based only on the insecurity of some men. Amīn recommended instead that

164 Ibid., 56.
165 Ibid., 29.
166 Ibid., 52.
167 Ibid., 53–4.
168 Ibid., 69, quoting the famous verse 24:30–1.
169 Ibid., 83.
170 Ibid., 65, 67–8.
171 Ibid., 103.

those men who fear seduction should avert their eyes (*yaghiḍḍ baṣarahu*) – which was in line with Islamic law.[172]

While the Quran mentions that women should cover their body in public, certain parts of the body are exempted.[173] Although not specified, these are probably the face and the hands, and possibly also the arms and feet. Amīn quoted several scholars, such as Ibn ʿĀbidīn, ash-Shāfiʿī, and ʿUthmān b. ʿAlī az-Zīlaʿī, who all agreed that at least the face and hands could be visible. Likewise, the Mālīkī and the Ḥanbalī theological schools believed that neither the face nor the hands belonged to a woman's *ʿawra* (literally, "nakedness," i. e., the private parts of the body).[174] Amīn argued in any case that the Quran and the *ḥadīths* allowed, but did not command, the woman to reveal her face and her hands, which, according to Amīn, would make sense to every reasonable person.[175]

Amīn went on to explain that the issue of seduction so prevalent in Islamic culture[176] stemmed not from the woman's body, but from the way that she walked. Amīn then reached the astonishing conclusion that, rather than decreasing a woman's power of seduction, the *niqāb* and the *burqaʿ* actually increased it! They were the most powerful means for a woman to provoke desire precisely since they hid a woman's identity. Only if her face were visible would it become clear to which family she belonged and she would be careful not to harm her reputation.[177] What sounds like an Orientalist fantasy was in fact written by the most popular and most influential advocate of women's liberation in the modern Arab world!

If the *ḥijāb* encourages bad manners, then it is only logical for the woman to cast off her veil. The woman, Amīn argued, was not and would never be able to complete her existence unless she took control of herself and enjoyed the freedom according to the law and the *faṭra*, so that she developed her talents to the highest degree possible. The *ḥijāb* would be a huge obstacle preventing the woman and, by extension, the *umma* from making progress. One of the most important reasons for the weakness of the *umma* lay in the fact that

172 Ibid., 77.

173 Quran 24:31: وَقُل لِّلْمُؤْمِنَاتِ يَغْضُضْنَ مِنْ أَبْصَارِهِنَّ وَيَحْفَظْنَ فُرُوجَهُنَّ وَلَا يُبْدِينَ زِينَتَهُنَّ إِلَّا مَا ظَهَرَ مِنْهَا "And say to the believing women that they should lower their gaze and guard their modesty; that they should not display their beauty and ornaments except what (must ordinarily) appear thereof" (Transl. by Yusuf Ali).

174 Amīn, *taḥrīr al-marʾa*, op. cit., 69 – 72.

175 Ibid., 76.

176 Assem Hefny, *Herrschaft und Islam*, op. cit., 74 – 5; cf. Kreutz, *Zwischen Religion und Politik*, op. cit., 195.

177 Amīn, *taḥrīr al-marʾa*, op. cit., 78.

women were excluded from employment, and that the education of children suffered from uneducated mothers. He also considered the mother's education to be impossible for as long as women were veiled.[178]

We can only understand this if we take the veil's context into perspective. "If we take a girl and teach her everything a boy learns in elementary school and convey good morals to her, then lock her up in the house and prevent her from having contact with men, there will be no doubt that she will forget gradually what she has learned." Knowledge needs to be put into practice, which is why "after a short while we will find no difference between her and an uneducated woman."[179] A bad education frustrates every *ḥijāb*.[180] Against the fear that bad morals might gain traction in society, Amīn argued that a woman who was unveiled and in contact with men was even less prone to bad thoughts because, once she was used to being in the society of men, she would be more immune to sexual desire.[181] Since it was unimaginable for a man to be imprisoned without having committed a crime, Amīn asked, then why was it reasonable for a woman to be held in custody and under the veil, so that she could not claim the virtue of her chastity (*ʿiffa*)?[182]

The decline of women's status in society was the greatest obstacle to progress.[183] The well-being of the *umma* required the well-being of the woman.[184] According to Amīn, a good education and an independent will were the two factors of human progress at any time and at any place. How, he asked, could the claim be made that both had a negative effect on women's souls?[185] A man who was willing to learn would understand that the woman's veil killed her personality.[186] Amīn emphasized once again that this was in full compliance with Islam, according to which the human was not created to be held by the reins, but had a natural disposition for science and knowledge.[187] From this perspective, a

---

178 فعند ذلك يرى [القارئ] أن المرأة لا تكون ولا يمكن أن تكون وجوداً تاماً إلا إذا ملكت نفسها وتمتعت بحريتها الممنوحة لها بمقتضى الشرع والفطرة معاً ونمت ملكاتها إلى أقصى درجة يمكنها أن تبلغها، ويرآ أن الحجاب على ما ألفناه مانع عظيم يحول بين المرأة وارتقائها وبذلك يحول بين الأمة وتقدمها ... من أكبر أسباب ضعف الأمة حرمانها من أعمال النساء وأن تربية الطفل لا تصلح إلا إذا كانت أمه مُربَّاة Amīn, *taḥrīr al-mar'a*, op. cit., 58.

179 Ibid., 86.

180 Ibid., 98.

181 Ibid., 93.

182 Ibid., 95.

183 Ibid., 127.

184 Ibid., 132.

185 فكيف يمكن لعامل أن يدعي أن لهدين العاملين ... حسن التربية واستقلال الإرادة هما العاملان في تقدم الرجال في كل زمان ومكان أثراً آخر سيئاً في أنفس النساء؟ Ibid., 101.

186 Ibid., 107.

187 Ibid., 120.

woman would not be a complete creature unless she were physically and intellectually educated – "physically," because a weak body can only be inhabited by a weak mind. That was why soundness of reason followed in every respect soundness of the body, which was an achievement that Amīn traced back to Anglophone civilization.[188]

The ascent of individualism in England was in fact accompanied by the decline of patriarchy.[189] It was against this background that the Maronite-Lebanese writer Aḥmad Fāris ash-Shidyāq (1804–87) explored the role of love and marriage in Western society in his book *as-sāq ʿalā s-sāq* (*Thigh over Thigh* (1855)).[190] The renowned Egyptian-Syrian author Jurjī Zaydān described in 1912 how Europe had overcome discrimination against women, and why it was possible for the Muslim world to embark on the same path. Zaydān theorized on the cultural differences between Egypt and Western Europe, where he had travelled to, in his 1912 book *Trip to Europe* (*riḥla ilā Ūrūbā*).[191] His account was devoid of any essentialism, and nor was he blind to the dark sides of European history. On the contrary, he made the reader aware of the changing role of women in Europe, where they had once been regarded as malicious and squalid, and had been treated like objects.[192]

But, as Zaydān went on to say, when the light of civilization (*tamdīn*) fell on the continent, outdated values and traditions were placed under the scrutiny of scientific knowledge. This paved the way for a re-evaluation of women's position in society, and finally for her improved status. For Zaydān, it was the growing awareness of women's importance to society and the family that led European women to demand their rights and to spark a controversy on the their future in society. In an apparent allusion to Delacroix's 1830 painting "La Liberté guidant le people," Zaydān's argumentation pointed to the fact that many European intellectuals made the woman a symbol of freedom and virtue. Zaydān therefore concluded that, of all the peoples of the world, the French respected women the most.[193]

More generally, he described in his itinerary the differences between Western and Eastern (Arabic) traditions, and how they resonated with the political situation of the respective countries. Beginning with the assumption that the structures and functions of the family differed according to place and time, he empha-

188 Ibid., 124.
189 Stone, "Der Wandel der Werte," op. cit., 352.
190 Lewis, "The Idea of Political Freedom," op. cit., 273–4.
191 Jurjī Zaydān, *riḥla ilā Ūrubbā sanat 1912* (Cairo: al-Hilal, 1923), 31.
192 Ibid., 45.
193 Ibid., 45–6.

sized that, until the end of the nineteenth century, the average Arab family resembled an absolute monarchy. A typical Arab family father used to rule his family members like a sultan ruled his subjects, and treated them as he saw fit. He told them what to do, where to live or travel, and whom to marry.[194] This tradition, as Zaydān pointed out, underwent radical change once the Arabs entered modern civilization (*madanīya hadītha*). Since that time, the contemporary family was organized more like a constitutional system, although, as Zaydān was forced to concede, the remnants of absolute rule had not yet been overcome.[195]

These ideas resonated with a general atmosphere of liberal thought in the Levant. Abolishing the veil became a demand of female emancipation, such as in Syria.[196] Among Turkish nationalists, it was Ziya Gökalp who celebrated the Turks as the oldest feminists among the nations of the world, which shows how much women's rights were an issue at that time.[197] In the 1930s, many Serbian students on the other side of the Mediterranean wanted to break out of the patriarchal role model, which is why they strove for relationships that were egalitarian. One of the new models was the "student marriage," which was sometimes concluded in the church, but without a wedding ceremony and often without the parents being invited. As Buchenau (2011) has shown, this model of marriage was based on the ideal of equality, but it failed when it came to working life, since most employers chose men over women.[198]

2.4 Towards a New Humanism

Qāsim Amīn's arguments should be seen against the backdrop of a time when a secular elite was beginning to develop its countries economically, while the clerics were stalling: "Nowadays it is our scholars' persuasion that devoting yourself to worldly matters and rational sciences means nothing to them."[199] This obstructiveness was not confined to the Arab world. In Iran, it was the reformist

---

194 Ibid., 37.

195 Ibid.

196 Dalal Arsuzi-Elamir, *Arabischer Nationalismus in Syrien: Zaki al-Arsuzi und die arabisch-nationale Bewegung an der Peripherie Alexandretta/Antakya 1930–1938* (Münster et al.: Lit, 2003), 128.

197 Cf. Ziya Gökalp, *The Principles fo Turkism. Transl. from the Turkish and annotated by Robert Devereux* (Leiden: Brill, 1968), 104–5, 111–2.

198 Buchenau, *Auf russischen Spuren*, op. cit., 350–1.

199 ومن رأى علمائنا اليوم أن الاشتغال بشؤون العالم والعلوم العقلية والمصالح الدنيوية شيء لا يعنيهم Amīn, *taḥrīr al-mar'a*, op. cit., 118.

cleric Shaykh Ebrāhīm Zanjānī who portrayed the situation in Iran in similarly gloomy terms: "As soon as you are able to talk and maybe to read and write a little, the only thing you learn from parents, neighbors, relationships, teachers, preachers, and poets about this world and the hereafter is that there is nothing more important than mourning the imams and visiting their graves. Especially in recent centuries it happened that every night and every day, every week and every year women, men, teachers, storytellers, preachers, authors, books, poems, meetings and celebrations gather around the funeral and spend their money while totally devoting themselves to that goal."[200]

This is similar to Qāsim Amīn's thinking. As a true Enlightenment figure, he reflected on the root cause of human misery, which he identified as "illusion" (*khayāl*). The more the human overcame his illusions, the closer he grew to happiness and truth. Provided by God, truth was what man had to search for, and was the only means for completing human reason and the human psyche. Both men and women needed to know the truth and to acquire reason to govern their souls, which was what took them to good deeds.[201] Then why have Muslims, why has humanity, been attached to illusion for so long? As Berlin (2007) once said, a preference for rewards in the next world resonates with a lack of individual liberty.[202] Individual liberty is often seen in the context of Islam as a threat to the *umma*. Many Muslim intellectuals such as the Syrian journalist and reformist al-Kawākibī in 1902,[203] have pondered on the reason for the decline of the *umma*, and found it in the decline of Islam.[204] According to Kawākibī, there are fundamental differences between Western and Eastern rule. For example, Westerners demand an oath from their ruler that he will be a loyal servant to the nation and abide by the law, while, in the East, it is the ruler who demands his subjects take an oath of allegiance to him, that they follow and show loyalty to him. Also, the Westerner is oriented towards the future and sincerity, while the Easterner is a son of the past and of fantasy (*khayāl*).[205] Such a sweeping judgement is of

---

200 Schaykh Ebrāhīm Zanjānī, *Khāṭerāt (sar-gozasht-e zendegī-ye man)*, ed. Gholāmḥossein Mīrzā Ṣāleḥ (Tehran: Melli, 2001), 3. Hat tip Dr Mahmoud Rambod, Bochum.

201 كلما تجرد الإنسان عن الأوهام والخيالات قرب من السعادة ويبعد :فإن كل مصائب الإنسان تأتي له من باب واحد وهو الخيال الحقيقة هي الكنز .الحقيقة هي ضالة الإنسان في العالم، ويجب عليه أن يسعى وراءها بلا قصور ولا تعب .عنها بقدر ما يبعد عن الحقيقة الحقيقة هي مشرق السعادة، لأنها الوسيلة وحدها .الذي أودع الله فيه كل آمال الإنسان، لا يجدها إلا من رغب فيها ومال عن سواها والنساء مثل الرجال في الحاجة إلى معرفة الحقيقة وإلى اكتساب عقل يحكم على نفوسهن .لوصول الإنسان إلى كمال العقل والنفس ويرشدهن في الحياة إلى أعمال الطيبة النافعة Amīn, *taḥrīr al-mar'a*, op. cit., 56–7.

202 Berlin, *Liberty*, op. cit., 207.

203 Barqāwī, *muḥāwala*, op. cit., 109.

204 Ibid., 93.

205 Ibid., 104–5.

course not tenable in its generalization, but it might nonetheless reflect different tendencies in different societies, and might therefore be not completely inaccurate.

One sort of illusion in the Islamic context is that Muslim thinkers often overemphasize the significance of their culture's impact on Western civilization. This supremacist thinking, which traces all Western values and achievements back to Islam, also assumes that all knowledge is already in the Quran and only needs to be extracted. Thus, the theologian Ḥusayn al-Jisr believed that the Christians in the West lagged behind due to their faith, and that the only way they could catch up with the level of Islamic civilization was to reflect on their religion and open themselves up to Islamic influence.[206]

The term "illusion" also played a significant role in the thoughts of Tawfīq al-Ḥakīm (d. 1987). In his play *ahl al-kahf* (*The People of the Cave*). which alludes to the legend of the Seven Sleepers. Ḥakīm focused on three main philosophical themes – love, time, and reality – to point out that there was a higher truth opposed to the truth of the human. The protagonists Mishlīnyā and Prīskā long for each other, but have to suffer from circumstances that they cannot control. Instead of accepting the truth, they long for an illusion that will never come true. For Ḥakīm, this symbolizes his idea that the divine is lost among people, which is the reason that they are in a state of confusion.[207] Ḥakīm also rejected economics for its ignorance of otherworldly matters, and thought that the prophets of the East advised people not to chase this-worldly matters for good reason,[208] since such matters were simply another example of "illusion."

According to Ismāʿīl Aḥmad Adham (1911–1940), the people of the East in general, and Muslim theologians in particular, both start out with God and end up with God. God is not only the maker of the world; he is absolute in his power and stands above the universe, which he sets in the correct order. God is the first and the final cause of everything that is and will ever be. The world is made in unity and harmony because it is subject to divine laws and rules, which never change in time. Man is tied to these laws and rules, and they constitute a permanent connection with God, whose will is inextricably

---

206 Al-Jisr, *Riyāḍ Ṭarābulus al-Shām*, op. cit., vol. 5 (1897): 30–2, quoted after Ebert, *Religion und Reform*, op. cit., 119.
207 بريسكا: (...) ما أجملك بطلاً من أبطال المآسي الإغريقية التي كنتُ أطالعها في خفية (...)، وأنا صغيرة Tawfīq al-Ḥakīm, *Ahl al-kahf* (Cairo: Dar ash-Shuruq, [1933] 2008), 3rd act, 71. Al-Ḥakīm on "illusion" in *ahl al-kahf*, v. Michael Kreutz, *Arabischer Humanismus in der Neuzeit* (Berlin: Lit, 2007), 95–8. On Ḥakīm's notion of human freedom v. Kreutz, *Arabischer Humanismus*, op. cit., 99.
208 On Tawfīq al-Ḥakīm v. Kreutz, *Arabischer Humanismus*, op. cit., 91–112.

linked to this system of being.[209] Contradicting al-Ḥakīm, he argues that we live right here and right now, and that we have to treat the world as if we were living in it forever. Adham saw another major difference between the West and the East in terms of affirming or denigrating the world.[210]

Born to a Muslim father and a Protestant mother of German descent, Adham grew up in Alexandria (which was then a marketplace for Western, and particularly French, cultural products) before becoming one of the few trailblazers of atheism in the Arabic world.[211] When he moved to Turkey in 1927, he founded the "Society for the Spreading of Atheism" (*jamāʿat nashr al-ilḥād*), which he later renamed the "Eastern Society for the Spreading of Atheism" (*al-majmaʿ ash-sharqī li-nashr al-ilḥād*), before continuing his way in 1931 to St Petersburg, where he studied mathematics until 1934. Adham was a multilingual scholar with a command of, amongst others, Turkish, Russian and German. His list of publications is sparse since he committed suicide at an early age.[212]

His most important work was a treatise called "Why I am an atheist" (*li-mādhā anā mulḥid*), which was influenced by Schopenhauer and Kierkegaard. It attracted little interest during his lifetime.[213] Adham had much in common with contemporary authors such as Ismāʿīl Maẓhar (1891–1962), editor of the Egyptian journal *ʿuṣūr* and a personal friend of Adham's who also tried to explain the social history of the Arab world according to Comte's "law of the three states" (*lois des trois états*). Comte (1798–1897) had already been *en*

209 الكون في التصرّف مطلق الخالق وانّ حادث العالَم انّ الى المسلمين، متكلّمة انتهى كما فانتهى، الخالق من بحثه بدأ الشرقي انّ ذلك
في المشهود التغاير عن البحث بينما ، سيكون وما يكون ما لكل والأخيرة الأولى والعلة يحدث ما لكل السبب وأنه له ومدبّر عنه منفصل
الغرب بمفكّري انتهى كما بالإنسان ينتهي كله وهذا والنظر، الاستنتاج اسلوب جانب إلى والمشاهدة الاستقراء بأساليب بالأخذ يدفع الكون
المكان، في ولا الزمان في لا تتغير لا ثابتة وسنن لنواميس خاضع وأنه وانسجامه، وحدته للعالم وأن الكون، في سببا حادثة لكل أن إلى
والضطرار اللزوم عنصر على قائمة وأفعاله الكون هذا بنظام مقيّدة الله إرادة بذلك وتصبح والنواميس، السنن بهذه قيّده الله إلى انتهى فإذا
Ismāʿīl Aḥmad Adham, "bayna al-gharb wa-sh-sharq" [=letter no. 260, June 27, 1938], in op. cit., 139–40. Cf. Wilber (1969): "In the daily life of a Moslem, the conscious toward the supernatural, or Allah, is continually expressed in the conventions of ordinary speech, general behaviour patterns, and attitudes toward natural events." Donald N. Wilber, *United Arab Republic Egypt: Its People, Its Society, Its Culture* (New Haven/Conn., 1969), 98, quoted after Soheir Taraman, *Kulturspezifik als Übersetzungsproblem: Phraseologismen in arabisch-deutscher Übersetzung* (Heidelberg: Julius Groos, 1986), 119.

210 Adham, "bayna al-gharb wa-sh-sharq," op c.it., 137.

211 Idem, "li-mādhā anā mulḥid," op. cit., 157. On his parent's background v. *muʿjam al-muʾallifīn*, ed. ʿUmar Riḍā Kaḥḥāla (Beirut s.a., 1993?):, Vol. 2, 285, lemma "Ismāʿīl Adham."

212 Adham, "bayna al-gharb wa-sh-sharq," op c.it., 158–9. About the circumstances of his death v. Ibrahim Abdel Meguid, *Birds of Amber* (Cairo: American University in Cairo Press, 2005), 218.

213 Abdel Meguid, op. cit., 282.

vogue among Egyptian intellectuals during the era of Muḥammad ʿAlī (1769–1849).[214]

For Adham, the Eastern mentality with its supposed inclination towards metaphysical thinking was the precise opposite to Western logic, which saw human life as it is and only attempted to analyze its social organization by means of reason. Using Comte's formula of social development, he located Western culture at the third, or top, level while he located the East at the second level, which is characterized by a blending of the perceptible and the metaphysical realm.[215] Comte described in his *Système de politique positive* (1851–1854) a pattern of general historical development of mankind. The first stage is marked by a theological or fictional worldview; the second, by a metaphysical stance that works as an intermediate stage; and the third, the positive stage in which all phenomena are explained through scientific laws. These stages correspond to the same number of epochs: the military and theocratic epoch is followed by the juridical, and finally by the scientific and industrial age, which Comte saw coming during his own lifetime.[216]

The West's starting-point in terms of thinking, Adham went on, is the search for difference in the world around us before we might finally take God into consideration. In contrast, the East's way of thinking aims at detecting the various forms of unity that then lead to God and to nature, which is the reason that philosophy first appeared in the West and theology first in the East.[217] People in the East might accept the existence of a "free will" (*irāda ḥurra*) at face value. However, even this "will" is subject to God's superior will (*irāda ʿulyā*). It is for this reason that fate plays a pivotal role in Middle Eastern thought, since it is God who determines history and whatever God has determined cannot be contested

214 Gershoni and Jankowski, *Egypt, Islam, and the Arabs*, op. cit., 114; Adham, "li-mādhā ana mulḥid," op. cit., 159; Kreutz, *Das Ende des levantinischen Zeitalters*, op. cit., 96.

215 المجموع افراد بين الصلات وتنظيم حقيقته معرفة يحاول وحده العقل طريق وعن هي، كما الإنسانية للحياة ينظر الغربي المنطق
وتنظيم الحياة تفسير يحاول الغيبي العنصر هذا طريق وعن الإنسانية، الحياة في غيبياً عنصراً يُدخل اللذي الشرقي الذهن بعكس البشري،
إلى انتمت وأنها إنسانية، الغربية الثقافة بأن كله هذا من نخلص ان ولنا .الإنسانية الهيئة افراد بين العلاقات وإقامة الانسانية الصلات
المرحلة حدود عند وقفت التي الشرقية الثقافة بعكس أوغست كونت، عنه كشف اللذي الإنساني التفكير مراحل من الأخيرة المرحلة
المنظور وراء ما بعالَم المنظور العالَم فيها يمتزج حيث الثانية، Adham "bayna al-gharb wa-sh-sharq," op. cit., 142.

216 Cf. Hourani, "Islamische Geschichte, Geschichte des Nahen und Mittleren Ostens, moderne Geschichte," in idem, *Der Islam im europäischen Denken* (Frankfurt/Main: S. Fischer, 1994): 109–41, here 112–3.

217 هنالك فرق اساسي في منطق التفكير بين الشرقي والغربي، هذا الفرق ينحصر في أن الشرقي يبدأ بحثه من الوحدة المتجلية حوله
فينتهي للخالق ومنه للطبيعة، بعكس الغربي الذي يبدأ بحثه من التغاير اللذي يكتنفه فينتهي للطبيعة ومنها للخالق Quote from an
older address, as stated by Adham.هذا الفرق المشهود في عن الشرقي يبدأ من عالَم الغيب لينتهي للعالَم المنظور، بعكس
الغربي اللذي يبدأ من العالَم المنظور وينتهي لعاللم الغيب، كان سبباً لظهور اللاهوت عند الشرقيين والفلسفة عند الغربيين Adham,
"bayna al-gharb wa-sh-sharq," op. cit., 139.

by man. If He says that something shall be, then it shall be.[218] Since, despite having a free will, the human cannot change what God has determined, he can only behave as though he had a choice, while in fact he is tied to the divine knowledge and the divine will, which is superior to every other will.[219] In Western eyes, the human, even when obeying the laws (*nawāmīs*) of life, does not give up on the idea of changing his environment in compliance with personal goals and future ambitions, even while making use of those laws of life for his own ends.[220]

According to Adham, humanity walks a path between the logic of the West and the spirit of the East. He stresses that the terms "West" and "East" do not simply denote geographical entities, but rather realms of mentality, which does not rule out the possible existence of a country in the heart of Europe whose way of thinking is similar to an Oriental way of thinking. We could also say that the medieval Europeans were Oriental in character, which was a side-effect of Christianity, which came from the Orient.[221] However, Adham rejects all sorts of biological or anthropological explanations, including the racist theories of Arthur Gobineau (1816–82).[222]

Adham also argued that Charles Martel's victory over the Arabs in the battles of Tours and Poitiers in 732 can be compared to the Greek victory over the Per-

218 هو الذي يقضي .والإنسان من حيث هو كائن في العالم المنظور، فهو في نظر الشرقي خاضع لإرادة عليا، هي إرادة الخالق الحرة
وهذه هي فكرة القضاء والقدر عند الشرقيين، فإذا قضى الله أمرا فلا مرد لقضائه، وإذا أراد شيئا قال له كُنْ فيكون .فيكون ويقدّر فيحدث
Adham, "bayna al-gharb wa-sh-sharq," op. cit., 140.

219 الإرادة؛ حرية الخالق وهبه الذي المخلوق الإنسان إرادة به تعلقت إذا إلا بوقوعه قُضي الذي بالأمر تتعلق لا الإلهية الإرادة أن غير
التي الإلهية وبالإرادة الأزلي الإلهي بالعلم مقيد الأمر فإن النظر عند أنه غير .اختياراً للإنسان فكأن بالأشياء الإنسان إرادة تتعلق فإن
إرادة أاي على ترجح Ibid.

220 عن له المقدّر تغيير قدرته في فإن لها، ويخضع الحياة نواميس وسلوكه تصرفاته في يتبع كان وإن فالإنسان الغربي نظر في أما
عن له، المقدّر وبين الوجود في ومطالبه الحياة في حاجاته بين مة الملاء إيجاد على والعمل بوجوده، المتحكّمة النواميس معرفة طريق
وصالحه يتكافأ بما تغييره طريق Ibid., 140.

221 وخلاصة القول إن في الشرق استسلاما محضا للغيب، وفي الغرب نضالا محضا مع قوى الغيب؛ وبين منطق الغرب وروح
إنّ ما نعنيه باصطلاح الشرق والغرب لا يقوم على أساسٍ من تقسيم العالم الى شرق (...) .الشرق تسير البشرية في قافلة الحياة
وغرب في تقويم البلدان، إنما ترجع التفرقة عندنا الى ما نلمسه من طابع ذهني للغرب ومنزع ثقافي للشرق؛ على اعتبار أن هذا
الطابع عام للغرب وذلك المنزع عام للشرق، غير أن هذا لا يمنع أن نجد مجتمعا غربيا ينزع منزع الذهن الشرقي في قلب أوروبا،
فمثلا يمكننا القول إن طابع التفكير في القرون الوسطى في أوروبا كان شرقيا في العموم، لغلبة .(...) في زمن من أزمنة التاريخ
Ibid., 140–1. المنزع الشرقي على الطابع الغربي نتيجة لبلوغ المنزع الشرقي شغاف أوروبا وغزوه الغرب مع الدين المسيحي

222 إن هذا المنزع الثقافي في الطابع الذهني لكل من الشرق والغرب، إذا اعتبرناه من الخصائص الأولية لشعوب الشرق والغرب،
فذلك لا يرجع إلى عوامل بيولوجية أو انتروبولوجية، كما حاول أن يثبتها بعض مفكري القرن التاسع عشر، بل يرجع إلى أسباب طارئة
اعلى المحيط الطبيعي والبيئة الاجتماعية؛ فلهذا لا يُرد علينا بما كتبه المناظر في الرد على غوبينو Ibid., 141. The same Gobineau also ordered the translation of Decartes' *Discourse* into the Persian which was released in 1862, v. Mohamed Tavakoli-Targhi, *Refashioning Iran: Orientalism, Occidentalism and Historiography* (Basingstoke and New York: Palgrave, 2001), 10; Mangol Bayat, *Iran's First Revolution: Shiism and the Constitutional Revolution of 1905–1909* (Oxford: Oxford University Press, 1991), 36.

sians, which he regards as having saved "the Western mentality from the tyranny of the Asian spirit of asceticism." Rationalism was at that time a burden on the shoulders of Christian theology, which made Rome a protector of the souls and minds while perpetuating all the negative elements of the Asian spirit of asceticism. Rationalism and asceticism began to clash when the idea of reformation arose and Luther paved the way for a new stage of intellectual renewal.[223] These words resemble a strong adoration for Protestantism, which, as we have seen, is not seldom among Muslim reformists.[224]

There have been attempts to overcome this "Oriental" or "Islamic" conception of the human. One such attempt was made by the Egyptian intellectual Salāma Mūsā Mūsā (1887–1958), an Egyptian Marxist of Coptic descent and author of a book about the Renaissance, for which he used the term 'Nahḍa'. Outlining why the Renaissance had been of crucial importance to the Arabs, Mūsā set up a typology with three stages of humanism in the West, starting with the first in Italy, the second in France, and the third with the appearance of Darwin.[225] The Renaissance promoted the idea that in this world humans must rely on themselves and work hard for their culture and happiness, since there is nothing in this world that they can rely on except reason.[226]

In this vein, the advancement of modern science happened at the expense of religious belief which, Mūsā went on to argue, was mostly associated with the Arabs. This is why Roger Bacon, the founder of scientific experiment, was accused of being a Muslim because at that time it was Muslims who were promoting science. Between 700 and 1300, the Arabs were the most advanced people in the world and, by being connected to the Mediterranean, they were open to the world.[227]

---

223 في ذلك الوقت كانت العقلية الغربية رازحة تحت كاهل اللاهوت الكنسي، الذي قام بروما رقيباً على النفوس والعقول مُحمَّلاً بكل غير أنّ العقلية الجرمانية لم تر في رقابة روما وتسلط البابا إلا روحاً آسيويةً بعيدة عن طبيعة الذهن ... سيئات روح التنسّك الآسيوية الغربي، فعملت كل الجهد في تقطيع أوصالها، وبدأ عهد الإصلاح بالصراع بين الذهنية الجرمانية الخالصة ممثِّلة العقلية الأوربية وبين في ذلك الوقت شقّ لوثر طريقه، وكان عصر الإصلاح الديني ... العقلية البابوية التي تحمل في طيّاتها شيئاً من روح النُسك الآسيوية وعهد الإحياء الفكري Adham, "bayna al-gharb wa-sh-sharq," op. cit., 145.

224 On the significance of protestantism for the islamic world as a potential engine against intellectual oppression v. Sharafī, *al-Islām wa-l-ḥurrīya*, op. cit., 165. For an overview of this debate v. Reinhard Schulze, "Islam und Judentum im Angesicht der Protestantisierung der Religionen im 19. Jahrhundert," in *Judaism, Christianity, and Islam in the Course of History: Exchange and Conflicts*, ed. Lothar Gall and Dietmar Willoweit (Munich: R. Oldenbourg, 2011): 139–65, *passim*.

225 Mūsā, *mā hiya an-nahḍa?*, op. cit., 34–5.

226 أي أن البشر، أو الإنسان، يجب أن يشتغل ويعتمد على ."النهضة لم تعنِ في الماضي، وهي لا تعني الآن شيئاً سوى "البشرية نفسه في هذا العالم ويعمل لحضارته وسعادته في جراءة وفهم. إذ ليس له في هذا الكون كله ما يعتمد عليه سوى عقله (...) Ibid., 35.

227 Ibid., 43–4.

Mūsā did not want to argue for an Islamic or Arab supremacy, but wanted to show instead how much European and Arab culture were intertwined, and how much both sides benefited from each other. He referred to Maimonides, the "Jewish-Egyptian philosopher" during the time of Saladin, as an example of someone who developed ideas through cultural connectedness. Therefore, the European Renaissance (*an-nahḍa al-ūrubīya*), which began in the fifteenth century, adopted four cultural achievements from the Arab world: the Indian numbers, the production of paper, the ancient Greek scriptures, and the scientific experiment.[228] Therefore, the Arab culture of science and the European the power of industry are akin because they both rest on empiricism.[229]

However, Mūsā viewed the Arabic-speaking *umma* of his time as lagging behind Europe because it had neglected science and industry, and was unable to restore its position in the "caravan of human advancement."[230] Likewise, it was not the church that was to be blamed for the decline of thought during the Middle Ages; rather, it was the human intellect itself that fell into decay when it began to see the world from a merely religious perspective and refrained from exercising doubt and investigation. This "ice age" of the intellect began in Alexandria in the late antiquity and gained momentum after the end of the Ptolemaic era, before coming to an end with the Arabs in Andalusia.[231]

Although the medieval era was famous for championing tradition over reason, Mūsā continued, it turned to reason in its final days. However, its thinking never went beyond the realm of religion, so that it transformed philosophy into theology. This, Mūsā goes on to explain, is the reason that we find three contradictory currents of thinking in the European Renaissance: the first, the return to the ancients in the arts, which is basically pagan; the second, the learning of everything to do with religion; the third, the scientific movement.[232]

Mūsā refers in his account to two medieval thinkers, Peter Abelard and Thomas Aquinas. He praises both for their desire to contribute to human culture (*ath-thaqāfa al-basharīya*) by learning from non-religious sources, and especially from pre-Christian literature. This is how they paved the way for humanism, which inherently reaches beyond a purely religious learning towards something entirely made by humans. However, they did not do away with religion, since studying human culture is likely to nurture spirituality. For Mūsā, the present

228 Ibid., 45.

229 Ibid., 48. Similarly Burhan Ghalioun [Burhān Ghaliyūn], *naqd as-siyāsa: ad-dawla wa-d-dīn* (Casablanca and Beirut: al-Markaz ath-thaqafa al-arabi, 2007), 245.

230 Mūsā, *mā hiya an-nahḍa?*, op. cit., 49.

231 Ibid., 51–2.

232 Ibid., 52–3.

is an echo of the Renaissance,[233] with its turning towards the human and its secular ideal of learning. It was therefore only consistent for Mūsā to praise Qāsim Amīn for his contribution to liberating women, because we cannot celebrate the human and ignore the lowly status of half the population.[234]

According to Mūsā, humanism[235] (*ḥarakāt basharīya*), the flipside of the Renaissance, is the view that this world is the goal of human activity and not anything that lies behind it, and so we have to rely on ourselves if we are to achieve happiness (*sa'āda*).[236] Mūsā points out that this was crucial for the Renaissance in Europe and had two tangible aspects. The first was based on history, religious critique, and the Greek and Roman arts, which constituted a literary movement as exemplified by Desiderius Erasmus (Erasmus al-Hūlandī, 1466–1536). The second was founded on science and empiricism, and fostered a rejection of traditions, or at least a lack of belief in their usefulness. Furthermore, humanism drew on invention and the search for scientific thinking driven by doubt and the wish to reform (*iṣlāḥ*). It therefore began a new path in human history and a novel approach towards nature. This second aspect was exemplified by the Italian artist and engineer, Da Vinci (1519–1452).[237]

From here, Mūsā begins to discuss what he calls the two main aspects of the Renaissance: the one built around literature, books, history, conversation, preaching, and study of the past, and the other dealing with inventions meant to improve the future of mankind. Science, rather than literature, was its vehicle.[238] The first found its geographical center in Italy and Germany, while the second began in France, where it morphed into a literary, social, political and religious renewal. These two Renaissances were completed by a third that began in the middle of the nineteenth century. Its main proponent was Darwin.[239] According to Mūsā, the first was at least partly a "religious Renaissance" (*nahḍa dīnīya*), and had taken place in Germany.[240] Luther's ideas of religious freedom were part of the Renaissance struggle for freedom of mind and led to freedom of conscience. This was made possible by an "autonomy of the human self" (*taqrīr al-maṣīr li-n-nafs al-insānī*), which meant that it was not priests or the church

233 Ibid., 54–5.
234 Ibid., 153.
235 Ibid., 101.
236 Ibid., 32.
237 Ibid., 65–6.
238 Ibid., 68.
239 Ibid., 105–6.
240 Ibid., 101.

that ruled over the human; rather, faith was something between the individual and God.[241]

England also witnessed a "scientific Renaissance" (*nahḍa ʿilmīya*), led by Bacon and Newton, as well as a literary Renaissance (*nahḍa adabīya*), shaped in France by Voltaire, Diderot and Rousseau, all of whom aimed at "the independence of the mind and the reliance on human thinking in light of the universe." Mūsā emphasized the significance of Newton, who was present at the beginning of the "modern industrial mechanical Renaissance,"[242] since "there is no civilization and no force without science." Mūsā stressed again and again the centrality of the scientific worldview. All of this was reflected by Diderot's encyclopaedia,[243] which collected and codified wisdom based on the idea that knowledge of the ancient was of no value in itself, but was nonetheless worth recording.[244]

As Blumenberg (2001) has pointed out, a major aim of every Enlightenment process and its notion of history was to make the consolidation of reason irreversible. Diderot's encyclopaedia became the foremost means to this end.[245] The era of the Enlightenment centerd on spreading secular knowledge and scrutinizing everything that claimed to be of divine origin. Driven by the desire to collect and make knowledge available to the masses, the eighteenth century was the era of lexicons and encyclopaedias.[246]

Mūsā views the Renaissance in his account as having triggered a "second Renaissance" driven by ideas shaped by Voltaire and Rousseau that sparked the Great Revolution in France (*ath-thawra al-kubrā fī Faransā*). Voltaire and his attacks on religions were the outcome of his study of ancient and modern religions, and his discovery of their secrets.[247] Mūsā praises Voltaire for advancing freedom of expression,[248] and the reliance of the human mind on the usefulness of the modern scientific spirit. It was this particular spirit that paved the way for free philosophical thinking (*at-tafkīr al-falsafī al-ḥurr*).[249]

---

241 Ibid., 56–7.
242 الاستقلال الذهني والاعتماد على التفكير البشري في مواجهة هذا الكون Ibid., 101–2.
243 Ibid., 156.
244 Ibid., 102.
245 Blumenberg, *Lebenszeit und Weltzeit*, op. cit., 202.
246 Hans Ulrich Gumbrecht, *Diesseits der Hermeneutik: Die Produktion von Präsenz* (Frankfurt/Main: Suhrkamp, 2004), 52.
247 Mūsā, *mā hiya an-nahḍa?*, op. cit., 103. وكانت نتيجة هذه النهضة، التي يمكن أن تُوصَف بأنها الحركة البشرية الثانية في أوربا، أن ثارت الثورة الكبرى في فرنسا وهي ثورة تجد فيها أثر فولتير في الدعوة إلى الذهن والمنطق وأثر روسو في الحملة على التقاليد والظلم Ibid.
248 Ibid., 154–5.
249 Ibid., 102.

Moreover, Mūsā recalls that this "autonomy of the human mind" already mentioned as being promoted by authors in Scotland and elsewhere included reading and reflecting on whatever people wished, be it the writings of ancient Greek or Roman infidels, and criticizing whatever issue they wished to criticize. Thus, "experience" became the new method of seeking the truth.[250] At this point, Mūsā connects the dots to emerge with some reflections on Arabic culture, starting out with an account of the great achievements of Arab culture and how the Arabs interacted with Europe. He portrays the Arabs in one chapter as being the source of the scientific mind set (*an-naz'a al-'ilmīya*), which was adopted by the West, making Western universities the gateways to Arab culture.[251]

Given the fact that great civilizations always engage in cultural exchange, there is little wonder that, with the consolidation of Islam, Baghdad became the center of Greek studies on Ptolemy, Archimedes, Euclides and Hippocrates, as well as of Indian studies, Mūsā goes on to explain.[252] The Jews played an important role here as facilitators of knowledge. In Spain they became a Westernized people (*mutagharrabūn*) when the Arabs left, so that they began to disseminate translations of the Greek sciences and the works of Arab scholars such as al-Khwārizmī, Ibn Sīnā and Ibn Rushd. They opened up medical schools and made use of Arabic books or of those translated from Arabic into Latin, sometimes via Hebrew. This development made possible the rise of Jewish scholars such as Ibrāhīm Bārshīyā, Ibrāhīm b. Ḥiyā (Ḥayyā), Rabbi Ibn Ezra and Mūsā b. Ṭibbūn. Moreover, the Arabs advanced Greek astronomy and thus laid the foundations for the European scholarly examination of the celestial sphere. Summing up, Mūsā concluded that the scientific Renaissance of Europe was based on empiricism, which the Arabs had a tendency towards. And that it was transmitted to Europe by the Jews.[253]

It was the areas of Greek science that Arabs were primarily interested in were those of medicine and mathematics. Yet, it was the translation of Greek philosophy via Arabic into Latin that caused religious people in Europe in the fourteenth century to study the books of the ancient Greeks in order to learn about rhetoric and religious disputation – issues that the Arabs showed no interest in.[254] Mūsā then turns to the Ottoman emperor Mehmet II and the conquest of Constantinople by the East Romans.[255] History, he explained, had cast a shadow

250 Ibid., 56.
251 Ibid., 95.
252 Ibid., 97.
253 Ibid., 98–100.
254 Ibid., 100.
255 Ibid., 121.

of Turkish ascendancy over our countries because this ascendancy was nothing but imperialism in the worst sense of the word.[256]

A repercussion of this development had been the emergence of humanism and the idea of free thought (*at-tafkīr al-ḥurr*), when Greek linguists were forced to go into Western exile and therefore contributed to the liberation of the European mind. The culture they brought with them became an incentive for the Europeans to expand into India and other regions of Asia, where Alexander the Great and Hellenism had set foot. From then on, the Europeans – or, more precisely, the Westerners – had ruled the world. It was the Turks who had unintentionally triggered the Renaissance in Europe.[257]

The new culture of "humanism" (*al-ḥaraka al-basharīya*) was based not on theology and religious books only, but rather on the "humane," i.e., on knowledge, not on creed. It was this movement that gave rise to science (*ʿilm*). Also, the Europeans realized that a pagan people such as the ancient Greeks did not know the meaning of religious fanaticism (*at-taʿaṣṣub ad-dīnī*). At that time, freedom of thought was permissible, and intellectuals wrote and spoke as if they had nothing to fear from state power.[258]

Turning to literature, Mūsā makes clear that the European folk literature of the medieval era was closer to the spirit of Eastern literature than to Latin and Greek literature, since the latter was made for the higher classes. This is why Eastern literature in its entirety is inclined to some form of fantasy (*khayāl*), depicted in dazzling bright colors. Europe had been heavily inspired by the Eastern spirit, so that it developed a strongly imaginative literature that it became so popular until it almost swept traditional literature away. With its scientific Renaissance, Europe embarked on studying Greek culture while neglecting the Eastern heritage, and modelled its literature on ancient Greek literature. Therefore, the ancient traditional tendency in literature largely replaced the imaginative tendency.[259]

However, the prevalent attitude in Europe was that Arab and Western literature were completely different because the latter had a basis in the Latin and Greek heritage. But a number of Orientalists and researchers had noticed that in fact it was Arab literature that was at the root of modern European literature. The most prominent scholar of this kind was the late HAR Gibb of the University of London. Both French and Italian poetry were strongly influenced by Arab po-

---

256 التاريخ قد ظلمنا باستيلاء الأتراك على أوطاننا، لأن هذا الاستيلاء كان استعمارٌ بكل ما تحمل هذه الكلمة من المعاني السيئة Ibid., 123.

257 Ibid., 126–7.

258 Ibid., 125.

259 Ibid., 89–90.

etry in Sicily, especially during the reign of the German emperor Frederick II. According to Mūsā, medieval European prose also owed a great deal to Arab prose.[260] Mūsā's message is that, as long as it was connected to the West, the East worked to liberate the Western imagination (*al-khayāl al-gharbī*) from its shackles and helped it to rid itself of the ancient traditional literature. Yet, Arab literature not only influenced its Western counterpart in style and method; it also had a moral impact, i.e., it transferred the spirit of the East (*rūḥ ash-sharq*).[261]

But the influence worked in both directions, since "the dignity of the industrial worker and his independence, then also his freedom of thought (*ḥurrīyatuhū l-fikrīya*), the equality of the sexes and respect for the constitution and the laws, all of these are the fruits of the industrial milieu and the milieu of the city, which stay away from the hardship of the village ... Finally, this freedom, both social and intellectual, which was unknown to the agrarian community, i.e., to an Eastern community (*umma sharqīya*) living on agriculture."[262]

Mūsā believs that the Renaissance was but the people's shift from being rural dwellers to urban dwellers, where commerce, handcraft and urbanization took place. For, where there is individual reason (*ra'y*) over faith, there is also the opportunity for exploration and invention. This thinking began to spread in Europe's cities in the course of the fifteenth century, when the new scientific method, as well as the close examination of the ancient Greek and Latin writings, emerged. The source of all this was the appearance of the merchant and the craftsman in the towns, and the expansion of Europe through navigation to other continents.[263]

Mūsā's view of the culture of Europe is ambivalent. Its downside was the ubiquitous phenomenon of anxiety (*qalaq*), stress (*tawattur*), and psychical problems. Yet, Mūsā remains loyal to Europe for its culture of learning, independence, democracy, science and invention – and, finally, of factories, which seemed to be of great importance for Mūsā, since "cannons are made in factories, not grown in fields."[264] He obviously admires the military strength of some European powers. In any case, he is not blind to the imperfect reality. Superstition and traditional interpretations of the world would not completely van-

260 Ibid., 87–9. On Gibb cf. Said, op. cit., 105–6.
261 Ibid., 92.
262 Ibid., 135.
263 Ibid., 62–3.
264 Ibid., 137.

ish, but material things demonstrated the power of science, based as it was on both experience and reasoning.[265]

Mūsā was not the only proponent of the Nahḍa to be fascinated by the European Renaissance. One of the most inventive discussions on that issue was that of the Egyptian writer Ṭāhā Ḥusayn.[266] As mentioned above, Ḥusayn argues repeatedly that ancient Greek democracy had put the idea of individual freedom into practice. We can find many remarks of this kind in his 1925 treatise on the *Pioneers of Thought* (*qādat al-fikr*). In *With Abū l-ʿAlā' in His Dungeon* (*maʿa Abī l-ʿalā' fī sijnihi*) (1939), he takes the life of the medieval Arabic philosopher Abū l-Ālā' al-Maʿarrī (d. 1058) as his starting-point for examining the different aspects of reason and their impact on individual life. Ḥusayn devotes himself to the question of how the individual is able to acquire knowledge beyond the scope of personal experience, and explores the extent to which human thinking can be autonomous, or whether it was a mere product of its environment.[267]

As a rationalist, Ḥusayn championes reason as a means of fighting illusion (*khayāl*).[268] In this regard, Abū al-ʿAlā' al-Maʿarrī seemed to be a credible forerunner, since he tried to evade illusion and be committed to reason alone, which is the reason that he attempted to live a life free of this-worldly distraction. He was driven by his commitment to a personal mission (*baʿth*). Yet, Ḥusayn is not blind to the downside of his thinking, since, in his retreat from the world, al-Maʿarrī appeared to have little interest in changing his environment – a lifestyle that stood in stark contrast to Ḥusayn's, since Husayn was a public intellectual and active politician.[269] Because al-Maʿarrī lived a life in complete isolation from other people, he ended up in a world of dreams where human action was nothing but an illusion – something that he was so keen to overcome.[270]

Why, then, did al-Maʿarrī not take this step and liberate himself from the deliberately chosen dungeon? Ḥusayn finds a possible explanation in the aforementioned "illusion," which, he claims, is so seductive to mankind, since it makes them believe that reason is something beyond reality. Thus, they keep on living in an isolated cave rather than finding their way out of it.[271] Here he argues against the hubris of any belief system that makes reason a replacement

265 Ibid., 134.
266 Cf. Kreutz, *Arabischer Humanismus*, op. cit., 97.
267 Ibid., 77–80.
268 Ibid., 78.
269 Ibid., 78–9.
270 Ibid., 82.
271 Ibid., 77–8.

ideology for liberty, when in fact its appropriate use is linked to the individual.[272] The lack of individual freedom is indeed a problem that is barely addressed by Levantine intellectuals.

A similar criticism of "illusion" or "imagination" was made on the other side of the Mediterranean, when Ionos Dragoumis lashed out against the centrality of the state in the thinking of the newly born Greek national movement: "The Greeks of Greece think that the Greek state is the reality. While the reality is everything else, is Hellenism, panhellenism, the people, the *ethnos*."[273] This criticism emerged from a nationalist perspective and distorted the ideas of the Enlightenment and humanism. When Dragoumis denies the reality of the Greek state as the center of Hellenism, he was denying, in his own words, any reality,[274] since reality "fought the constitution! The constitution sought to give the nation representatives and said there are deputies. The MPs were to represent the nation. But the reality spoiled everything and the MPs found themselves representatives of a small province and had no interest in being engaged in the nation."[275] Reflecting on the state of Hellenism, Dragoumis turns to the situation in Egypt, where, for him, the life of the Greeks was "better than in any other place." Since the conflict between Greeks and Ottomans had ceased,[276] the time had come for "the second or third blossoming (*anthisma*) of Hellenism in Egypt."[277]

---

272 Ibid., 55–6, 62, 77.

273 "Οἱ Ἕλληνες τῆς Ἑλλάδος νομίζουν ὅτι τό ἑλληνικό κράτος εἶναι ἡ πραγματικότης. Ἐνῶ ἡ πραγματικότης εἶναι κάθε ἄλλο, εἶναι ὁ Ἑλληνισμός, τό Πανελλήνιον, τό Γένος, τό Ἔθνος. [...]" Dragoumis, Ὁ Ἑλληνισμός μου καί οἱ Ἕλληνες, op. cit., chap. 3, § 4: 49–50.

274 "Ἀρνοῦμαι τήν πραγματικότητα τοῦ ἑλληνικοῦ κράτος ὡς κέντρου τοῦ Ἑλληνισμοῦ, ὅπως ἀρνοῦμαι κάθε πραγματικότητα. Πραγματικότητας εἶναι ἡ φαντασία μου." Ibid., chap. Γ, α Athens 1905, § 6: 50–1.

275 "Τό σύνταγμα τό πολέμησε ἡ πραγματικότητα! Τό σύνταγμα θέλησε νά δώση στό ἔθνος ἀντιπροσώπους καί εἶπε νά ὑπάρχουν βουλευτές. Ἐπρόκειτο αὐτοί οἱ βουλευτές νά ἀντιπροσωπεύουν τό ἔθνος. Ἀλλά ἡ πραγματικότητα τά χάλασε ὅλα καί οἱ βουλευτές βρέθηκαν ἀντιπρόσωποι μιᾶς μικρῆς ἐπαρχίας καί δέν εἶχαν κανένα συμφέρον νά καταγίνουν μέ τό ἔθνος. [...]" Ibid., 135–6.

276 "Γίνεται φυσιολογικό φαινόμενο ἡ ζωή τῶν Ἑλλήνων στήν Αἴγυπτο καλλίτερα παρά σέ κάθε ἄλλο μέρος. [...] Στήν Αἴγυπτο οἱ κοινότητες πῆραν τή φυσιολογική τους μορφή· εἶναι φανερά ἑλληνικές κοινότητες. [...] Οἱ κοινότητες ἔτσι εἶναι παραρτήματα τῆς Ἑλλάδος· παύει ἡ πάλη Ἑλλήνων ὑπηκόων καί ὀθωμανῶν ὑπηκόων." ibid., chap. Γ, β Alexandria 1905, § 2: 59, no. 3.

277 "Τώρα εἶναι καί τό δεύτερο ἤ τρίτο ἄνθισμα τοῦ Ἑλληνισμοῦ στήν Ἄιγυπτο [...]. Iibid., chap. Γ, β Alexandria 1905, § 2: 61, no. 5.

2.5 From Turkey to Japan

Egypt was a welcome place for those Greeks who found life too distressing in Asia Minor and not sufficiently commercial in Greece. This is why they left for Egypt, "where they flourish, spread and open up" in line with the Greek "commercial, profit-oriented character."[278] The disposition towards commerce is central to Dragoumis' description of the Greeks, who left their country twice (after 1453 and above all after 1821), before fleeing to the West, where they freely continued their trade with Vienna, Pest, Trieste, Odessa, Moscow, Liverpool, London, Marseille and Venice.[279] Reformist intellectuals in the former territories of the Ottoman Empire – with the sole exception of Albania – had aimed since the nineteenth century to modernize their societies by overcoming all remnants of the Turkish (i.e., Ottoman) past.[280] The effect was far from resounding in the Balkans, which is why Todorova (2003) argues that it is the Ottoman heritage that has caused a lasting cultural tension between Western Europe and Western Asia.[281] Some intellectuals like the Serbian geographer Jovan Cvijić or the aforementioned Ionos Dragoumis propagated a Eurasian cultural sphere characterized by its European, Western Asian and Levantine patrimony.[282]

This resonates strongly on the other side of the Levant, where a political author such as Sulaymān al-Bustānī viewed the Ottoman Empire as being characterized by its spanning of three continents. From Sudan to Iran to the Adriatic, writes Bustānī, "you can see our country being tied to Asia, Europe and Africa by natural ties, which makes it especially strong." Moreover, this was where commerce had been strong since ancient times, so that the Empire was tied to a cer-

---

278 "Βρίσκουν στενόχωρη τή ζωή στήν Τουρκίαν οἱ Ἕλληνες ἤ ὄχι ἀρκετά ἐμπορική στήν Ἑλλάδα καί γι' αὐτό φεύγουν ἀπ' ἐκεῖ καί ἔρχονται ἐδῶ στήν Αἴγυπτο καί ἐδῶ ἀνθίζουν καί ἐκτείνονται καί ξαναοίγουν καί πλαταίνουν καί εἶναι ζωντανοί καί εἶναι σύμφωνοι μέ τόν χαρακτῆρα μας τόν εμπορικό, τόν κερδοσκοπικό." Dragoumis, Ὁ Ἑλληνισμός μου καί οἱ Ἕλληνες, op. cit., chap. Γ, β Alexandria 1905, § 2: 60, no. 4.

279 "Ἐπί τουρκοκρατίας οἱ Ἕλληνες (μετά τό 1453 καί πρό πάντων μετά τό 1821) ἔφυγαν πρός τήν Δύση, καί ἐκεῖ ἔκαμαν τό ἐμπόριό τους ἐλεύθερα (Βιέννη, Πέστη, Τεργέστη, Ὀδέσσα, Μόσχα, Λίβερπουλ, Λονδῖνο, Μασσαλία, Βενετία)." Ibid., chap. Γ, β Alexandria 1905, § 2: 60, no. 5.

280 Clayer, "Der Balkan, Europa und der Islam," op. cit., 317.

281 Cf. Maria Todorova, "Historische Vermächtnisse als Analysekategorie: Der Fall Südosteuropa," in *Wieser Enzyklopädie des Europäischen Ostens*, Vol. 11: *Europa und die Grenzen im Kopf*, ed. Karl Kaser, Dagmar Gramshammer-Kohl and Robert Pichler (Klagenfurt, Vienna and Ljubljana: Wieser, 2003): 227–52, here 226; cf. Clayer, "Der Balkan, Europa und der Islam," op. cit., 326.

282 Clayer, op. cit., 318.

tain degree to the spirit of antiquity.[283] The "natural ties" that he referres to were the people, "both in the past as in the present, who were great in commerce and who travelled all over the world," exploring other cultures and connecting them – something that is supposed to characterize the Arabs in particular.[284]

This is a line of argumentation similar to that of Dragoumis' (see 99). Bustānī goes on to say that commerce fostered transportation projects such as the Suez Canal in order to promote its own acceleration.[285] It also helped to change the political and legal background when, for example, a treaty of commerce was signed in 1838 between England and the Ottoman Empire in order to facilitate the import of goods.[286] Bustānī imagines a future where all kinds of commerce thrived and industry connected East and West until the country became the axis of world commerce.[287]

This looks like a concession to Ottoman rule, but it is not. More than their fellow Muslim countrymen, the Christians tended to view Ottoman rule as something alien to their culture, which is at least one reason why they felt a kind of proximity to the West and why they championed stronger commercial ties to Europe. The feeling of uprootedness was especially strong among Arab Christians in the Levant. Historically, the increase in commerce was accompanied by a process of urbanization that saw educated people moving to the seaports, and especially to Beirut and Alexandria, which were better connected to Europe and later on to America.[288]

It was against this backdrop that the Maronite Synod founded a modern educational system (see p. 49) that would become the medium of modern ideas from Western Europe. As von Grunebaum (1956) has pointed out, the question behind modern-era Westernization was: "Can we ever become the political

283 ان بلادا كبلاد الدولة العثمانية، وهي عروة الوصل بين قارات العالم القديم الثلاث، كان يجب ان تكون قابضة على اوثق ازمة التجارة. ففحيثما سرحت نظرك على موقعها في رسم الكرة، من ضفة الطونا الى السودان ومن بلاد ايران الى بحر الادرياتيك، رأيتها مرتبطة بآسيا واوروبا وأفريقيا بصلات طبيعية تجعل لها ميزة خاصة تعز على ما سواها (...) Bustānī, *ʿibra wa-dhikrā*, op. cit., 218.

284 فحيثما استتبّت لهم (...) وان في هذه البلاد شعوبا كان لها في كل زمان قيدما وحديثا شأن في التجارة عظيم، يطوف ابناؤها البحار فستعمرون استعمار الفينيقيين او يستطلعون استطلاع العرب الذين بلغوا (...) قوة اليد والمال زادوا على زمرة تجارهم المقيمين في البلاد بطوافهم اطراف العالم الجديد. Ibid., 218.

285 ولكن في تحويل (...) .ان التجارة في البلاد العثمانية جارية بمجراها الطبيعي بمعنى انها تقوى وتضعف بالعوامل الطارئة عليها من جهة الى اخرى، او فتح طرق جديدة خراب بلاد وعمار بلاد، كما جرى بعد فتح ترعة السويس، اذ تحولت جميع [للتوسع] هذه الطرق يكون للتجارة حركة لم تكن في اللحسبان (...) .تجارة العجم وبعض تجارة الهند الى هذا الطريق Ibid., 219.

286 Weber, *Die jüdische Gemeinde im Damaskus des 19. Jahrhunderts*, op. cit., 47–8.

287 ونظرتُ بعين البصيرة الى المستقبل، فتصورت ما يكون من شأن جميع هذه الاصناف بعد بسط العدل واستتباب الامن وتسهيل وسائل النقل برا وبحرا، واذا أضفتُ الى ما تقدم تناج الصناعة المقبلة مع ما تعلمه من اتساع هذه البلاد وتوسطها بين الشرق والغرب، يخيل لك انه لا يطول بها العهد حتى تصبح المحور الاعظم لتجارة العالم. Bustānī, *ʿibra wa-dhikrā*, op. cit., 220.

288 Sharabi, *al-muthaqqafūn al-ʿarab wa-l-gharb*, op. cit., 15.

equals of the West without Westernizing ourselves completely?"[289] From this angle, the modern Arab Renaissance, the Nahḍa, was driven by the idea that the cultural roots of Western Europe thrived on the Arab heritage; hence, East and West are no longer different entities but contributors to one and the same civilization.[290]

However, as Barqāwī (1988) reminds us, it is no fewer than four centuries that separate the European Renaissance from the Arab Renaissance of the nineteenth and early twentieth century. If the European Renaissance marked the end of the medieval era, then the Arabic Renaissance marked the beginning of the disintegration of the Ottoman Empire as well as the shattering of its economic and cultural foundations at a time when Western Europe was shifting to imperialism. Therefore, the term "Nahḍa" points to a time of transition accompanied by a growing anger against feudalism, an increasing national consciousness, and a call for political freedom, all of which were influenced by ideas of the French Enlightenment. Although the Nahḍa was predominantly a secular phenomenon, the simultaneous emergence of an Islamic religious reform movement was part of the wider picture.[291]

In general, Western ideas were seen as an intellectual means to eliminate everything Turkish (i.e., Ottoman), so that even defenders of Arab nationalism agreed to defend Europe.[292] The Eastern feudalist Ottoman state as an autocratic Islamic entity with its predominantly Turkish element[293] constituted the background for the deteriorating image of the Turks in the nineteenth and twentieth century.[294] Qāsim Amīn (1863–1908) argues in 1899 that both the Turks and the

289 Hanf, "Die christlichen Gemeinschaften," op. cit., 34–5; Von Grunebaum, "The Problem of Cultural Influence," op. cit., 95–6.

290 بل الغرب هو الشرق في .في هذه الحالة ـ حالة اكتشاف الغرب في التراث العربي ـ لم يعد الغرب بالنسبة إلى النهضوي غربياً
مرحلة سابقة من تقدمه وازدهاره Barqāwī, *muḥāwala*, op. cit., 44.

291 Ibid., 15–6; cf. 31. فإن ... حسبنا القول أن أربعة قرون ـ تقريباً ـ تفصل عصر النهضة الأوربية عن عصر النهضة العربية
وإذا كان عصر النهضة الإوربية هو عصر بداية .تاريخ النهضة العربية يبدأ من القرن التاسع عشر وينتهي في بداية القرن العشرين
تحطيم ما شادته العصور الوسطى أي بداية نشوء العلاقات الرأسمالية، فإن تاريخ النهضة العربية يشير إلى بداية انحلال الامبراطورية
إذاً فالنهضة كمفهوم يشير إلى .العثمانية وبداية تحطيم أسسها الاقتصادية والثقافية في عصر انتقال الرأسمالية الأوربية إلى امبرالية
ولقد اجْمَعَ كثير من الباحثين ـ عرباً كانوا أم أجانب ـ على اعتبار ظهور حركة تنويرية في الشرق العربي، ... مرحلة انتقالية
وبعث الآداب العربية، واتساع المزاج المعادي للأقطاع، وظهور ونمو الوعي القومي، والدعوة إلى الاستقلال السياسي، ونشوء
أجمَعَ الباحثون على اعتبار .. حركة الاصلاح الديني الإسلامي، والتأثير القوي والحاسم من قبل الثقافة الغربية على الثقافة العربية
هذا كله نهضة عربية يبدأ من القرن التاسع عشر l Ibid., 15–6.

292 Ibid., 70. قد قوى الميل إلى تصور (...) من الصعب أن لا نعترف أن الارتباط بين الحركة القومية في بدايتها وبين الفئة المثقفة
ولهذا ترافق الدفاع عن قومية عربية مع الدفاع عن أوربا .أوربا، ذروة الكمال الحضاري، والنظر إليها كمخلص من الاستعباد التركي
وبخاصة فرنسا وانكلترا

293 Ibid., 67.

294 Haarmann, "Ideology and history," op. cit., *passim*, esp. 186–9.

Egyptians were united in "being ignorant, lazy and on the decline." But, as long as there was nothing that united all Muslims except their religion, Europeans and even a large proportion of the Muslim elite would conclude that religion was the sole reason (*as-sabab al-waḥīḍ*) for the decline of the Muslims and their backwardness. However, Amīn remains an apologist for Islam by stating that "the true Islamic religion" (*ad-dīn al-islāmī al-ḥaqīqī*) was surely not the reason for the decline of the Muslims, since the achievements of the past had proved that religion was a noble means and a powerful factor for taking mankind to the path of progress and to the goal of happiness (*sa'āda*). But what Muslims nowadays call religion supposedly encompasses "a variety of fancied creeds, practices and manners which have nothing to do with the pure and true religion but are inventions (*bida'*) and novelties attached to it." Amīn criticizes this "hodgepodge" of persuasions that the masses wrongly take for Islam and that he saw as being the true obstacle to progress.[295]

These intellectual developments must be understood vis-à-vis the predominance of the Turkish element, which fanned Arab aspirations for a new political order calling for either political independence or at least some reform leading to a non-centralist state.[296] An article in the journal *al-Hilāl* (1914) on social structure in Palestine presents the Turkish men of government (*rijāl al-ḥukūma al-atrāk*) and their entourage as constituting the *khāṣṣa*, the upper class, while the *'āmma*, the lower class, consisted mainly of the local Arab population, often working as civil servants, as well as the Christians and Jews, i.e., the *ahl adh-dhimma*, who stood at the bottom of society.[297] Thus the normal grievance against the privileged role of the Turks came as no surprise.

As the article goes on, this was already the social system when Bonaparte had conquered Egypt and Syria by the end of the eighteenth century, bringing with him the seeds of modern civilization. When new schools were founded, they had a missionary purpose, and it was therefore Middle Eastern Christians who were first to be attracted to these schools and their education. They were eager to learn the customs of the Westerners (*ifranj*), and especially of the French, so that the "spirit of the age built on individual freedom" became popular among them. The significance of foreign languages quickly rose in the field of commerce. This had a lasting impact on the Muslim population, so that the brightest among them began to acquire Western knowledge, too. Consequently,

295 Amīn, *taḥrīr al-mar'a*, op. cit., 115.
296 مما ولّد احساساً قومياً عربياً وحركاتٍ سياسية قومية تدعو إما إلى الاستقلال السياسي .اضطهاد واضح من قبل العنصر التركي الكامل أو إلى اصلاح سياسي أساسه دولة لا مركزية Barqāwī, *muḥāwala*, op. cit., 20.
297 "Filasṭīn: tārīkhuhā wa-āthāruhā wa-sā'ir aḥwālihā al-ijtimā'īya wa-l-iqtiṣādīya wa-l-'ilmīya," Part 3: *aḥwāluhā al-ijtimā'īya*, in *al-Hilāl*, Vol. XXII, No. 7, April 1914: 513–607, here 513.

there emerged in Syria and Palestine a new class of educated people consisting of both Christians and Muslims.[298]

Similarly, Bustānī stresses the importance of personal freedom, which is manifest in the freedom of the press, the "mouthpiece of the nation," as a means to popularize ideas of reform and to point to the flaws requiring attention. Therefore, the basic law facilitated the freedom of research and criticism, but, in the case of the Ottoman Empire, it was "turned into a trumpet of glorification and a demon of intimidation."[299] As a consequence, many newspapers were shut down or suspended for some time only because they had quoted some items of news from European papers. Usually it was harmless news about, for example, the murder of a minister in China or a prince in Africa, or even something unpolitical such as the invention of a new piece of technology. The editors were then surprised to learn that their newspaper had been suspended and spent months trying to discover why – without receiving any other explanation than that it was "required."[300]

As a result, the editors became eager to avoid anything that could lead to their paper's suspension. Therefore, only a few papers continued to operate, and those that did offered little more than some brief items of world news, a few articles about internal affairs and the good deeds of the government or the appointment of a governor or an officer.[301] According to Bustānī, journalists were prevented from publishing in their own newspapers and the censorship finally affected newspapers from all over the world, and particularly what was published in Egypt, which, according to Bustānī, was "the country most loyal to the Ottoman government and to the Ottoman people."[302]

Many journals were very difficult to obtain, such as *al-mu'ayyad* and *al-liwā'*; and, when they were allowed to be published, they were difficult to find on the markets outside the Ottoman lands. The two journals were bullied only because they repeatedly mentioned buzzwords such as freedom, constitution, in-

---

298 Ibid., 514.

299 واذا كان هذا شأن الحرية الشخصية، فما عسى ان يكون شأن حرية الصحافة تلك الآلة الحية الناطقة بلسان الامة المنبهة الافكار فانه وان كان القانون الاساسي قد اطلق سراحها على ما اتسع .المرشدة الى الاصلاح، المشيرة الى مواطن الخلل المنادية بحي على الفلاح يوسع لها في حرية البحث والنقد، فقد اصبحت بعد ذلك تحت مراقبة حولتها . له وقتئذ وأنشيء لها نظام مخصوص حوالي سنة ١٢٨١ هـ الى أبواق تمجيد وأغوال تهديد Bustānī, *'ibra wa-dhikrā*, op. cit., 97.

300 فكم من جريدة الغيث او اوقفت لزمن محدود لخبر روته عن جرائد اوروبا ينبئ بمقتل وزير في الصين او امير في افريقيا، او وظل صاحبها يبحث "بتعطيلها" بل كم من مرة فاجأ الجريدة الامر .اختراع ذكرته لآلة تطير في الهواء او غواصة تسير تحت الماء "الايجاب" اشهرا فلا يعلم لذلك سببا غير. Ibid., 97.

301 Ibid., 98.

302 ولكن .إذ لو حرمت علينا الكتابة في جرائدنا وابيحت لنا قراءة الصحف المنتشرة في سائر الاقطار لقلنا شر اهون من شرين ... حظرت المراقبة قراءة كثير من الجرائد المنتشرة في كل بلاد الله ولا سيما ما صدر في مصر اصدق البلاد ولاء للخلافة .هيهات الاسلامية والامة العثمانية Ibid., 99.

dependence, parliament, and similar notions that aroused feelings among the masses and rage in the minds of the ruling class.[303] Bustānī recalls that he and his colleagues had tried to operate within accepted borders, but did not wish to give up on their desire to express their own thoughts and views on domestic affairs, to draw their own conclusions in terms of a helpful lesson, and to offer a new perspective. They aimed to carry out research on everything enjoyable, educative and useful. It is interesting to learn that Bustānī wants to prevent freedom from morphing into something like chaos, which had supposedly happened to some newspapers in Egypt years before.[304]

As Bustānī explains, it was not so much tyranny in the sense of an absolute reign but rather an unjust rule that made people suffer from crimes and the breaking of "taboos." The rule of scoundrels, Bustānī writes, made the people bow down their heads and finally their souls. Tyranny was driven by pig-headedness and complete arbitrariness, and lacked checks and balances. Criminals could assault innocent people without fear of punishment. Tyranny ruled by dividing people and making them fight against each other. This distracted them from the worrying political situation in which a tiny group controlled all the wealth and indulged in luxury without ever being monitored.[305]

Journalists were unprotected by any rule of law. If someone examined a source in order to make use of it and told the government that the law allowed him to do so, the ruling class would ignore that person's right. Likewise, if someone presented a project as the fruit of long research and as something that he him-

---

303 فهل عرفت قبلهما او بعدهما صحيفة اشد تمسكا بالعرش العثماني واعظم تفانيا في خدمته، فهل اتيح لهما ارسال جريدتيهما الى كل هذا لان اللواء والمؤيد يرددان على صفحاتهما ذكر الحرية والدستور والاستقلال ... البلاد العثمانية مع ما فيها من كثرة طلابهما؟ في عرف رجال المابين "وتخدش الاذهان" والمجلس النيابي وما اشبه من الالفاظ التي تنبّه الشعور في عُرْف الناس Ibid., 100.

304 واننا بلا ريب لا نطمع ولا نود ان نتخطى الآن الى ما وراء المعقول، فنثِب وثْبة واحدة من وهْدة المسكنة الاضطرارية الى قمة التهور الاختياري بل جل ما نتمناه ان تباح لنا رواية الاخبار وترديد صَدَى الافكار والنظر في شؤون انفسنا من القاء درس مفيد وعرْض وعلى الجملة، اطلاق الحرية الى ما لا يفضى بها الى مثل .مقترح جديد ونقد عامل وعمل والبحث في كل ما من شأنه أن يلذ ويهذب ويفيد وهو لا شك ما ينظر اليه دعاة الدستور من الآن بعين الروية والتدبير .الفوضى التي استحكمت بين بعض جرائد مصر لسنين مضت Ibid., 102–3.

305 فمعظم الشكوى اذاً ليست من الاستبداد بمعنى الحكم المطلق، وان كانت دولة هذا الحكم قد دالت، وانما هي من ذلك الاستبداد، بمعنى الحكم الجائز، الذي اباح الموبقات واستباح المحرمات – استبداد حكم الانذال برقاب الرجال فنكس الرؤوس وذلك النفوس – ... ولا شرع له ولا وازع يحلل اليوم ما يحرمه غدا .استبداد لا مرشد له الا التعنت عن هوى تميل به النفس الى حيث لا تدري اذا انس نقمة من الناس عليه عمد الى التفريق بينهم فأثار فيهم ثائرة التعصب الذميم، .ويبطش المجرمون بالابرياء بغير عقاب فضرب بعضهم ببعض، حتى اذا غفلوا عن مظالمه حينا ثم استفاقوا من غفلتهم ورجعوا الى التظلم منه، خلق لهم ملهاة اخرى اذا ... .ولا حرج على تلك الفئة ولا جناح .يلتهون بها عنه – استبداد تقتسم فيه فئة ضئيلة أموال الامة فتتنعم بها وتشقى الامة اكتشف مجتهد منجما وقال للحكومة أنا صاحب الحق باستخراجه فلكم سهمكم ولي سهمي بمقتضى النظام، قال رجال المابين، بل هذا نتاج بحث طويل، ولدي جميع الوسائل :واذا قضى باحث زمنا فدرس مشروعا وقال .هو هبة استوهبها احدنا فاذهب خاسرا تلك هي الفئة الظالمة .فأخذوه بلا شرط ولا بدل .بل هو لنا، :العلمية والمالية للقيام به بهذه الشروط وذلك السهم منه للحكومة، قالوا وهو الذي احرج صدور العثمانيين فسهل لهم .ذلك هو الاستبداد الذي نقصده في بحثنا ... .التي كانت تتسبب بالنفي والسجن والقتل المنية في سبيل الحرية، حتى اذا نالوها بجهاد جيشهم الباسل ودعاتهم الاماثل، تصاعد صدى حماسهم فخرق لب الاثير Ibid., 88–9.

self had funded completely, then the government would tell him that it was all theirs and take it without compensation. It was the same tyrannical class that promoted deportation, jail and murder. This was the general political situation that made the Ottoman people so angry and fanned their longing for freedom.[306]

This general anger vented against Turkish, i.e., Ottoman, rule as described by Bustānī peaked in a pamphlet that was published by the nationalist *Comité arabe révolutionnaire* in 1914. The pamphlet combines emotiveness with self-victimization and an ostentatious hatred for the Turks.[307] The pamphlet begins with a call to the Arab masses to rise against the Turks (Ottomans), whose culture is regarded as inferior to that of the Arabs and as a culture that has brought nothing but mayhem and devastation.[308] The Turks are also held responsible for the fall of Andalusia into the hands of the Franks after the Ottoman sultan refused to shore up his Arab co-religionists. Also, according to the pamphlet, it was the Turks (Ottomans) who sank the civilization of Baghdad into the waters of the Euphrates and Tigris, as they were responsible for the fall of Tunisia and Algeria into the hands of the Europeans, which is why hundreds of thousands of Arab co-religionists had to die. Finally, the Young Turks stand accused of having ignored the Italian conquest of Libya's Tripoli.[309]

Similarly, Salāma Mūsā also claims that the Arab decline was the result of Turkish (Ottoman) destructiveness. When the Arabs were engaged in cultural exchange with Europe, they almost had their own Renaissance, since that was the mode of working between the two civilizations, i.e., to convey and to receive. However, the Turkish occupation had prevented this development. In Mūsā's narrative, the Arabs had remained in isolation for three centuries before Napoleon turned the situation upside-down and made them resume their connections with Europe as well as with modern culture. This was something that they had never obtained from the Turks, who had never handed over a rising culture (*thaqāfa nāhiḍa*). This is in contrast to the Indians when they adopted the English language and culture from the British emperors so as to become a modern nation.[310] The Arabs instead were caught in a medieval culture, preferring creed

---

306 Ibid., 89.

307 FO 618/3 [Kew, London] Comité Arabe Révolutionnaire (*ath-Jam'īya al-thawrīya al-'arabīya*), Translation, 27 June 1919.

308 FO 618/3 [Kew, London] Damascus, 2 May 1914, General Report for March Quarter.

309 FO 618/3 [Kew, London] 27 June 1919, Comité Arabe Révolutionnaire – *al-jam'īya al-thawrīya al-arabīya*. Cf. Kreutz, *Das Ende des levantinischen Zeitalters*, op. cit., 265.

310 واحتجنا إلى قرابة .ولكن الاحتلال التركي حال دون ذلك .وهو اتصال كان جديراً بأن ينقل إلينا نهضتها .وكنا على اتصال بأوربا ثم لم نكسب من الأتراك لغة حية .ثلاثة قرون، ونحن في عزلة، إلى أن جاءنا نابليون فشرعنا نستأنف اتصالنا بأوربا والحضارة العصرية

over knowledge, and old over new. However, due to global trade between Asia and Europe, which has found two points of contact in Cairo and Alexandria,[311] the light of the new is dawning.[312]

Mūsā's line of argumentation struck a popular chord at a time when the Arab East was trying to catch up with the West.[313] This endeavour was more than simply looking back to see what had gone wrong; it was also looking forward to search for a positive role model. The model that they found was not a Western power. In fact, reformists like al-Afghānī and many other Eastern intellectuals were fascinated by the rise of Japan, which was perceived as both an Eastern power and a modern state.[314] When Ziya Gökalp (1876–1924), the trailblazer of a modern Turkish nationalism, devoted himself in his attempt to make Turkey a modern and European country to the idea of becoming modern without giving up one's own culture or religion, he drew on the idea of civilization. Gökalp saw civilization as a concept that was independent of geography and ancestry, and he argued that the Japanese proved how a nation could be part of European civilization without being located in Europe.[315]

Qāsim Amīn took an Arab view on Japan: "We have seen in this century an astonishing incident which I regard as something unique in history: we saw a nation completely relinquishing its habits, nullifying its regulations, ridding itself of its customs and laws, and throwing them behind its back." Amīn was thrilled by the idea that a nation proved able to break with most of its ties to the past, and could set up new structures to replace the old. Within half a century, a concept for modernizing society had been established and brought to fruition, a concept injecting "strong and young blood into its veins." Is that, he wondered, not a lesson for everyone?[316]

أو ثقافة ناهضة كما كسب الهنود مثلاً من الإنجليز، حين أخذوا بلغتهم وثقافتهم اللتين جعلتا منهم أمة عصرية Mūsā, *mā hiya an-nahḍa?*, op. cit., 123–4.

311 Ibid., 123.

312 قد الجديد الفجر نور ولكن .الجديد على والقديم المعارف على العقائد نؤثر .الوسطى القرون ثقافة في نحيا بعيد حد إلى زلنا ما أنا بزغ Ibid., 130.

313 فالتأثير إذاً .فالشرق العربي المتخلف وجد نفسه وجهاً لوجه أمام الغرب المتقدم، والزاحف نحوه لتوظيفه في خدمة زيادة ثروته مارسه متقدم على متخلف Barqāwī, *muḥāwala*, op. cit., 43.

314 Ibid., 56.

315 Kreutz, *Das Ende des levantinischen Zeitalters*, op. cit., 185; Gökalp, *The Principles of Turkism*, op. cit., 39. On the CUP's idea of establishing a "Japan of the Middle East," v. Feroz Ahmad, *The Making of Modern Turkey* (London and New York: Routledge, 1994), 39.

316 رأينا في هذا القرن حادثة عجيبة أظنها وحيدة في التاريخ، رأينا أمة بتمامها خلعت عوائدها وأبطلت رسومها وتخلت عن نظاماتها وقوانينها وطرحتها وراء ظهرها، فقطعت كل صلة بينها وبين ماضيها، إلا ما كان متعلقاً بجامعة شعبها، ثم همت فبنت بناءً جديداً مكان البناء القديم، فلم يمضِ عليها نصف قرن إلا وقد شيدت هيكلاً جميلاً على آخر طرز أفاده التمدن، فهبت من نومها، ونشطت من عقالها، وشعرت بأن الحياة تدب في بدنها وتجري في عروقها دماً حاراً قوياً فتياً، تلك هي الأمة اليابانية، صارت تعد اليوم في صف الأمم

Admiration for Japan reached its peak in 1904/05, when a major event occurred that had a lasting impact on how the country was perceived. This was the Russian-Japanese war, which broke out when Russia invaded Korea and infringed the Japanese sphere of influence. It was the first war in which an Asian power ended victorious over a European power, and was given great attention on both shores of the Levant. Japan's victory would not have been possible without its advanced economic state, which was accompanied by an expansionist policy. Russia's defeat was generally evaluated according to an Asia *vs.* Europe pattern, which is the reason that so many people in the Islamic world identified themselves with Japan. According to documents from the British foreign ministry, the dividing line was a sectarian one: Muslims sided with Japan, and Christians with Russia. Japan was viewed from a Muslim perspective as an Ottoman Empire of the Far East.[317]

The Japanese victory was greeted with great applause by Arab intellectuals. In 1909, Muḥammad Kurd ʿAlī, editor of the journal *al-Muqtabas*, who argued that reform and renewal were necessary to halt the alleged decline of Arabic culture,[318] called for an "Arab-Western culture." This he already saw as emerging in both Cairo and – Japan. Instead of simply adopting Western concepts, he favored a synthesis, and Japan seemed to be a model that proved that amalgamating Eastern and Western culture was possible and worth adopting.[319]

This reflects a widespread enthusiasm for Japan beyond the Arab world, with, for example, ʿAbdarraḥmān Talibov, a pioneer of Iranian reformism, translating the Japanese constitution into Persian.[320] The Iranian reformist Seyyed Ḥasan Taqīzādeh noted that the "Russo-Japanese War in 1904 and the subse-

---

المتمدنة بعد أن قهرت في بضعة أيام دولة الصين الجسيمة التي لم يقتلها إلا إعجابها بماضيها، أليس في ذلك عبرة لكل متبصر؟ Amīn, *taḥrīr al-mar'a*, op. cit., 182–3.

317 Kreutz, *Das Ende des levantinischen Zeitalters*, op. cit., 174–5. FO 618/3 [Kew], Archives Damascus, 1904, 23–4: "The interest taken place here in the Russo-Japanese war is intense, the sympathies of all the Moslemo-Turko, Arabs, Kurds and Circassians – being unmistakably on the side of Japan while the Christians with very few exceptions are fervently praying for the success of Russia. [...]." This is confirmed by French archival sources, viz. Athènes, série A, no. 245 [Nantes]: Analyses de Presse et analyses de revues, 1905.

318 Sharabi, *al-muthaqqafūn al-ʿarab wa-l-gharb*, op. cit., 5; Samir M. Seikaly, "Damascene Intellectual Life in the Opening Years of the 20th Century: Muhammad Kurd 'Ali and *Al-Muqtabas*," in *Intellectual Life in the Arab East, 1890–1939*, ed. Marwan R. Buheiry (Beirut: American University of Beirut, 1981): 125–53, here 127.

319 Kreutz, *Das Ende des levantinischen Zeitalters*, op. cit., 182; Seikaly, "Damascene Intellectual Life," op. cit., 137–8.

320 Abdul-Hadi Hairi [ʿAbdolhādī Hā'erī], "The Idea of Constitutionalism in Persian Literature prior to the 1906 Revolution," in *Akten des VII. Kongresses für Arabistik u. Islamwissenschaft*, ed. Albert Dietrich (Göttingen: Vandenhoeck und Ruprecht, 1976): 189–207, here 203.

quent defeat of Russia made a deep impression upon the minds of the Iranians. The Revolution that followed in Russia after this defeat spread to the Caucasus, the immediate neighbor of Iran in the north, and greatly influenced the growing underground fermentation." These events finally led to the revolution of 1906 and the establishment of a democratic parliamentary government.[321]

Therefore, Japan and its constitution became a symbol of modernity.[322] Salāma Mūsā was also fascinated by Japan, this "age-old Asian nation" that had not reached a global level of development before 1914, when it began to embrace the "principles of European culture." Mūsā makes it abundantly clear what that means for the Arab nations: "I can't imagine a contemporary Nahḍa for any Eastern nation if it is not based on the European principles of freedom, equality and constitution, together with the scientific-objective view of the universe."[323] Like Gökalp, Mūsā presupposed a difference between culture (*thaqāfa*) and civilization (*ḥaḍāra*).[324]

As a consequence, the Arabs can adopt the European culture of progress without thereby relinquishing their own civilization. The Americans are also Westerners although they are not Europeans – while in the Western countries there are also Arab, Indian and Chinese people who have adopted European manners, as well as its culture and lifestyle, and who have embraced the constitutional and civil systems of law, i.e., parliamentary rule, equality between the sexes, the objective view of this world, and the collective feeling of responsibility for the individual.[325]

In Mūsā's view, civilizations, while transcending continents and individuals, are able to assimilate cultures other than their own. This also holds true for the Arabs and the Turks, who for a long time suffered under the "Ottoman occupation," which Mūsā compares to the supposed darkness of the European Middle Ages or even worse. But, when the Ottomans conquered Constantinople and caused the Greek intellectuals to emigrate (*hijra*) to Europe, they triggered a process whose repercussions would finally benefit the Turks:[326] Atatürk made the

321 Seyyed Ḥasan Taqīzādeh, "The History of Modern Iran (Lectures given in Columbia University), in idem, *Opera Minora: S.H. Taqīzādeh's Articles and Essays*, Vol. VIII: *Unpublished Writings (in European Languages)*, ed. Iraj Afshar (Tehran: Shekufan 1979), 209.

322 Fukuyama, *Political Order and Political Decay*, op. cit., 345.

323 Mūsā, *mā hiya an-nahḍa?*, op. cit., 132.

324 Ibid., 139.

325 Ibid., 131.

326 Mūsā, *mā hiya an-nahḍa?*, op. cit., 124.

people rise before leading them into the twentieth century and to what Mūsā calls "Atatürk's Renaissance" (*nahḍat Atātūrk*).[327]

Ziya Gökalp harboured similar ideas regarding Japan. "The Jews and the Japanese," he argued in his 1923 *Principles of Turkism*, "are alien to Europeans both in culture and religion, yet they have the same civilization as European nations."[328] For Gökalp, the Japanese had entered Western civilization at the expense of their former Far Eastern civilization.[329] For this reason, they are considered a European nation, he argues, although they have never given up on their religion or national identity, while the Turks still hesitate to take the same step.[330] Gökalp was convinced that, although it abided by the "old crusader fanaticism," i.e., occupying Muslim territories, Europe, had changed substantially insofar as it no longer defined itself by religion."[331]

In contrast, the Egyptian atheist reformist Ismāʿīl Adham viewed Japan as a rather negative example of modernization. Adham believed that it was impossible for Arabic society to benefit from positivist science without adopting its logical foundations. In this regard, the Japanese seemed to prove the theory that Eastern nations tended to make use of Western culture only as a means for their own technical advancement, and did not internalize its deeper intellectual foundations for the sake of knowledge. According to Adham, the Japanese refused to adopt the European logic of thinking (*manṭiq al-tafkīr al-ūrūbī*) and instead stuck to their Orientalness (*sharqīya*), and to their traditional pattern of thought. Here, he points to the reforms that had been introduced during the 1867–8 *Meji* Restoration by the so-called "Enlightened Government," reforms with which Japan embarked on a unique path to prosperity.[332]

It was against this backdrop that Adham criticized Japan for its allegedly superficial tackling of modernity. In his view, Japan's achievements were of a merely derivative character, since they were the outcome of another culture transplanted to Japan without this culture ever taking roots. He also addressed the fact that eighty million people lived in poverty, which he did not expect to change as long as the Japanese acquired no sense for human life and dignity. A case in point was Europe, which had freed itself from the yoke of pure ration-

327 Ibid., 128–9.
328 Gökalp, *The Principles of Turkism*, op. cit., 39.
329 Ibid., 40.
330 Ibid., 47.
331 Ibid., 64.
332 Cf. Akira Iriye, "Japan's drive to great-power status," in *The Emergence of Meiji Japan*, ed. Marius B. Jansen (Cambridge: Cambridge University Press, 1995): 268–329, *passim*.

alism.[333] This, indeed, is a very Kantian thought ("Kritik der reinen Vernunft"), and utters the popular view of Islam as a religion of reason.

What Western Europe does differently, according to Adham, is that it places life on a human basis and leaves it to science to work out how to organize human societies; in contrast, the East places life on a metaphysical basis and leaves the same process to metaphysics. Where the West harmonizes the needs of the individual and its environment by means of science, the East places life on a basis of mutual confidence.[334] Adham makes an important point here since the modernized Japanese state would also became the basis for an authoritarian fascist regime similar to the one in Germany – both constitute examples of the "dialectics of Enlightenment."[335]

Germany also played a pivotal role in respect to the Balkans. The Japanese victory and the settlement with Britain in Central Asia made Russia concentrate its imperial ambitions on the Balkans, where a clash with Austro-Hungary and Germany would become inevitable.[336] A similar enthusiasm for Japan was voiced in the 1930s by the Belgrade-based Balkan Institute (*Balkanski Institut*) before it was closed by the Nazis after Germany invaded the Balkans in 1941. Founded by the journalist Ratko Parezanin, the Institute called for a modernization of the region according to the model of Japan.[337]

333 Adham, "bayna al-gharb wa-sh-sharq," op. cit., 137;, cf. Kreutz, *Das Ende des levantinischen Zeitalters*, op. cit., 174–5.

334 Ibid., 137.

335 Cf. Fukuyama, *Political Order and Political Decay*, op. cit., 198–9.

336 Christopher Clark, *Sleepwalkers: How Europe Went to War in 1914* (London: Penguin, 2013), 159.

337 Buchenau, *Auf russischen Spuren*, op. cit., 369–70.

3 Civilizations Drifting Apart

3.1 The Past and the Future

The Protestant theologian Ernst Troeltsch (1865–1923) once noted that Europe is not founded on a mere perception of antiquity but[1] constitutes a dynamic concept[2] that made intellectuals on both shores of the Mediterranean eager to forge their nations in a European spirit. Especially in the Arab world, intellectuals viewed the ancient Greek and Roman heritage as a vehicle for progress and as a key to connecting with Europe so as to overcome the perceived backwardness caused by the Ottoman occupation.

The idea of cultural entanglement, particularly in terms of ancient Greek heritage, became prevalent among secular intellectuals at that time. Of course, the ancient Greeks were no "Europeans" in the modern sense of the word.[3] Lord Cromer, Britain's consul general in Egypt (1883–1907), often drew on Roman history to condemn the Islamic present.[4] Ismāʿīl Adham was thrilled that Arabic-Islamic philosophy had Greek origins, with al-Fārābī (ca. 870–950), Ibn Sīnā (ca. 980–1037) and Ibn Rushd (d. 1098) being distant from any Orientalism (*sharqīyat ar-rūḥ*) and basing their ideas on Greek philosophy and logic so that he also calls those local philosophers who tied God's will to the laws of being and of the universe as Europeans (i.e. Greek) in thinking.[5]

There was similar enthusiasm for the Pharaonic era. An Egyptian modernizer and reformer such as Rifāʿa Rāfiʿ aṭ-Ṭahṭāwī is a good example of this attitude, since he devoted himself to writing an historical work in three volumes, which was published in 1868 under the title *anwār tawfīq al-jalīl fī akhbār Miṣr wa-tawthīq banī Ismāʿīl* ["Highlights of Tawfiq the Great in the History of Egypt and the

---

1 Ernst Troeltsch, *Der Historismus und seine Probleme. Erstes (einziges) Buch: Das logische Problem der Geschichtsphilosophie* [2nd reprint of the ed. Tübingen 1922] (Aalen: Scientia Verlag, 1977), 716.

2 Ibid., 704.

3 Cf. Peter Burke, "Did Europe exist before 1700?," in *History of European Ideas*, Vol. I (1980): 21–9, *passim*, esp. 22–3, 25. "It looks as if by the year 1700 Europeans were more ready to talk about Europe, to see it as a whole, and to contrast it with the rest of the world than they had been in 1500, let alone the Middle Ages. Consciousness of being European was now an important social and political fact." Ibid., 26.

4 Toner, *Homer's Turk*, op. cit., 189.

5 Adham, "bayna al-gharb wa-sh-sharq," op. cit., 141. إن الفلسفة الإسلامية التي ظهرت على يد الفارابي وابن سينا وابن رشد، وغيرهم من أعلام الفلسفة الإسلامية، ليست شرقية الروح، بل وليدة الفلسفة اليونانية والمنطق اليوناني ويمكنك بكل سهولة أن تعود بخطوط فلسفة فلاسفة الإسلام إلى أصولها عن طريق أفلاطون وأرسطو وفلاسفة الإسكندرية من أفلوطينيين؛ فمن هنا لا ايُعترَض علينا بأن هنالك من الفلاسفة الشرقيين من علقوا إرادة الخالق بسنن الوجود وقوانين الكون

https://doi.org/10.1515/9783110634006-005

Confirmation of the Descendants of Ismail"]. In this book, he depicts his Egyptian fatherland from the Pharaonic era to the present day as being an arena of different civilizations.[6] Lewis (1998) has called this an "epoch-making book," one that had "added several thousand years to what the Egyptians knew about their own history."[7]Like Ṭahṭāwī, the great Egyptian reformer Ṭāhā Ḥusayn describes Egypt in his 1925 treatise *qādat al-fikr* [*The pioneers of Thought*] as a country that was deeply influenced by a pre-Islamic Greek culture and as a country therefore tied to the history of Europe.[8] Many authors were keen to find evidence of the ancient Egyptian origins of modern achievements, and even of the Enlightenment, which, according to Salāma Mūsā, for example, was rooted in Pharaonic Egypt. As his fellow patriot Muḥammad Ḥusayn Haykal has pointed out, Egypt had always been the spearhead of history and the center of the world, a site of the eternal battle between the forces of light and darkness, truth and falsehood; its mission was therefore an eternal one.[9]

In terms of ancient culture, reference was made also to Semitic peoples. An article dealing with "The vestiges of the Semites in European civilization, or: the vestiges of the Orient in the civilization of the Occident" that was published in Jurjī Zaydān's journal *al-Hilāl* (1914) stated that there was little wonder that the East and West were entangled, since both were adjacent civilizations and, if they had not been divided by the sea, would have constituted a single cultural entity. The article goes on to say that the main agents of exchange between East and West were Phoenician travellers, although there was some friction (*iḥtikāk*) between the Romans and Carthaginian Phoenicians. Other agents of exchange included Greek scholars who obtained knowledge from Egypt and other countries; Jews who left their homes due to oppression, for the sake of commerce, or in order to spread their religion; Arabs conquering and colonizing Spain, France, Sicily and Italy; the Crusaders during the medieval period; and, finally, "Arab Jews" (*al-yahūd al-ʿarab*), i.e., Jews from the Arab lands who had settled in Europe after the disappearance of the Andalusian state.[10]

---

6 Kreutz, *Arabischer Humanismus*, op. cit., 26.

7 Bernard Lewis, *The Multiple Identities of the Middle East* (New York: Schocken Books, 1998), 69.

8 Kreutz, *Arabischer Humanismus*, op. cit., 55.

9 Gershoni / Jankowski, *Egypt, Islam, and the Arabs*, op. cit., 181–2.

10 N.N., "āthār as-Sāmiyīn fī l-madīna l-ūrubīya aw āthār ash-sharq fī madīnat al-gharb," in *al-Hilāl* no. 6, March 1, 1914: 403–15, here 403–6. The Semites are defined as. "the Arabs and their brothers among those peoples communicating in semitic languages such as Arabic, Hebrew, Syriac, Phoenician, Assyrian and others."

The article in *al-Hilāl* emphasizes that Christianity and Judaism, the dominant denominations in Europe and America, are both thought to be Semitic in origin.[11] The European mentality is therefore Semitic in character, which is the reason that Europeans think in terms of notions provided by the Torah and the Gospels, and everything else that is Semitic in origin.[12] The social system of European civilization as far as Christians are concerned is also built on the teachings of the church, which in turn is also based on the Gospels and the Torah.[13] Jewish life is based on the Torah, the Talmud, and everything Semitic in origin. It is for this reason that, "if you witness a wedding party in Paris, London or Berlin, you will find that it does not differ essentially from wedding parties in any village in Lebanon."[14]

According to *al-Hilāl*, the Arabs conveyed many elements of Semitic culture to the Europeans when they settled in Spain. They had a great impact on the *ifranj* (lit. Franks) with regard to their daily lives. The Spaniards were therefore the first to adopt elements of Arabic culture, which, the article goes on to state, can be seen in a variety of fields, including language and mythology. Thus, many of the legends (*aqāṣīṣ*) recited in, for example, France, Italy, Spain, Germany and Great Britain can be traced back to Wadi an-Nīl and Arabia. Europe also owes games such as chess, backgammon, football and polo to the Semitic peoples.[15]

The article also argues that, in terms of ethics (*akhlāq*) and manners (*ādāb*), there are many similarities, too. This is because European customs were adopted from the Greek, Roman and Germanic peoples, and acquired a fairly Semitic coloring when they were passed on to the Europeans via Christianity and Judaism. The pagan-era ethics of Romans and Greeks differed from those of the Christian era of today, which in turn is mostly homogenous, so that the ethics of the French Christians are identical to those of the Christians of Syria, Palestine and Egypt.[16] However, the Semitic roots are all too obvious, since the Romans took from the Greeks, and the Greeks took from the Phoenicians.[17] According to this narrative, the "Franks" were influenced by the achievements of Semitic

11 Ibid., 405.

12 فعقول الاوربيين ونفوسهم عليها طابع سامي اصبح جزءاً من شعورهم Ibid., 405

13 فنظام الاجتماع في مدنية اوربا عند النصارى مبني على تعاليم الكنيسة واساسه الانجيل والتوراة Ibid.

14 ولو شهدت حفلة زواج في باريس او لندن او برلين رأيتها تختلف في جوهرها عن حفلة من نوعها في قرية من قرى لبنان Ibid., 406.

15 Ibid., 406–7.

16 Ibid., 407.

17 Ibid., 408.

culture in terms both of science and literature. This became especially apparent in the Renaissance, when modern European civilization was founded.[18]

Nevertheless, as the article concedes, cultural influence was not only unidirectional. Indeed, European manners and customs had a strong impact on Arabic clothing and food, and European literary styles swept (albeit unintentionally) into the writings of contemporary Arabic authors.[19] The article goes on to state that, since in terms of geography the Arabs were the pioneers of navigation, commerce and conquest, their knowledge was useful to European discoverers such as Marco Polo and Vasco da Gama. Indeed, when Roger II, ruler of twelfth-century Sicily, wanted to collect information on his realm, the only scholar that he could find who was suitable for this task was the Arab geographer ash-Sharīf al-Idrīsī. Hired by Roger II, al-Idrīsī drew up the *Tabula Rogerina* [*kitāb nuzhat al-mushtāq fī ikhtirāq al-āfāq*], a map of the world that would become popular in Europe for a long period of time.[20]

Enthusiasm for the Phoenicians was particularly strong in Lebanon, where the political discourse became increasingly oriented towards Western Europe and was occasionally embedded in a cultural-geographical framework established by French authors such as Maurice Barrès and Elisée Reclus. This line of thought was promoted by the historian Henri Lammens (1862–1937) in his popular book *La Syrie: Précis historique*, which became an essential source for many advocates of Lebanese nationalism. Lammens' concept of cultural continuity draws on the ideas of the French nationalist Maurice Barrès, and locates the roots of greater Syria (including Lebanon) in the culture of the Phoenicians, which he saw as being pivotal to the culture of Europe. The philosophy of Maurice Barrès (1862–1923), which emerged in nineteenth-century France, has been labelled "romantic nationalism." In his 1888 book *Le Culte du Moi*, he outlines a political theology based on Catholicism as the core of French culture, and on France as the guardian of Western civilization. This concept was palatable to Lebanese Maronite nationalists.[21]

---

18 نعني ما اقتبسه الافرنج من علوم العرب وآدابهم في نهضتهم الاخيرة لانشاء تمدنهم الحديث Ibid.

19 والاساليب الافرنجية تتسرب الى اقلام كتّابنا وهم لا يشعرون او لا يريدون .فترى العادات الاوربية تتطرق الى ملابسنا واطْعِمتنا. وترى بين شعرائنا من يقلد الافرنج في الشعر المنثوراو .على انهم كثيراً ما يتعمدون تحدي الافرنج في ما يعجبهم من آدابهم وطرقهم الشعر الموزون غير المقفي مما لا يوجد في آدابنا العربية Ibid., 414.

20 Ibid., 415.

21 Kreutz, *Das Ende des levantinischen Zeitalters*, op. cit., 136–7, 162; Asher Kaufman, "Phoenicianism: The Formation of an Identity in Lebanon in 1920," in *Middle Eastern Studies*, Vol. 37, No. 1, January 2001: 173–94, here 178, 183–5, 188–9, cf. Matl, op. cit., 277; idem, *Reviving Phoenicia: in Search for Identity in Lebanon* (London: Tauris, 2004), 25–6.; Kamāl aṣ-Ṣalībī, *Tārīkh Lubnān al-ḥadīth* (Beirut: Dar an-Nahar, 1991), 157–8; idem, "Islam and Syria in the Writings

An identical observation had led the Egyptian philosopher Zakī Najīb Maḥmūd (1905–1993) to muse on how the Arabic and the Greek had parted ways. For Maḥmūd, the Greek spirit was shaped by a different geography and had to deal with limitations to human endeavour such as mountains and forests, while the Arabic spirit was shaped by the vastness of their landscape, which facilitated a vision of infinity.[22] Maḥmūd is certainly right in that the Arabian Peninsula is marked by huge spacious grounds, less urbanization and less population density. Yet, Islam is not a religion of the desert.[23] A similar thought was expressed in a 1916 article in *al-Hilāl* in which the unknown writer sees a deep connection between the landscape of the Greek peninsula and the mentality of its people. In his concluding thoughts, the writer argues for the superiority of the Greeks regarding their societal systems, constitutions (*dasātīr*), and their "love for travelling and colonization."[24]

The Egyptian author Salāma Mūsā once emphasized that the terms 'European' and 'Western' in our contemporary minds are not just understood geographically, but supposedly reflect the older term 'Hellenic', which refers to the Greek people at a time when Greek culture was spreading and becoming prevalent. The word 'Hellenic' was therefore used to denote a certain tendency, philosophy and style, which made it possible for Egyptians and other Arabs to call themselves 'Hellenic'.[25] 'Europe', usually deduced from the Semitic *urūb* or *ghurūb*, is founded on Arabic heritage. Mūsā lists a number of words in Arabic, Greek and Latin to demonstrate the cultural entanglement between East and West.[26] Mūsā depicts the Syrian town of Palmyra (Tadmur), which was an important center of political power and cultural influence during the late antiquity, as an "Arabic-Greek state."[27] He then deals with the development of the democratic ideal. Understood as the "system of society," democracy means the rule of the people, i.e., the people rule themselves. In ancient Greece, democratic rule took place within the

of Henri Lammens," in *Historians of the Middle East*, ed. P.M. Holt, Bernard Lewis and William M. Watt (London: Oxford University Press, 1962): 330–42, *passim.*

22 Moustafa Maher, "Umrisse einer neuen Kulturphilosophie in Ägypten seit dem 19. Jahrhundert," in *Gegenwart als Geschichte. Islamwissenschaftliche Studien: Fritz Steppat zum fünfundsechzigsten Geburtstag,* ed. Axel Havemann and Baber Johansen (Leiden et al.: Brill, 1988): 309–18, here 317–8.

23 Tilman Nagel, *Angst vor Allah? Auseinandersetzungen mit dem Islam* (Berlin: Duncker und Humblot, 2014), 66.

24 N.N., "al-Yūnān majduhā al-māḍī inbiʿāthuhā l-ḥadīth," in *al-Hilāl*, vol. XXIV, No. 4, Jan. 1916,: 268–77, here 270.

25 Mūsā, *mā hiya an-nahḍa?*, op. cit., 131.

26 Ibid., 117–9.

27 Ibid., 117.

framework of small entities called *poleis*, which were the forerunners of modern democracies in Europe and America.[28] In its spread to other parts of the world, "the tree of democracy" was growing.[29]

Mūsā argues, however, that governing oneself requires some degree of education. Since education was not compulsory in England before 1870, there was no democracy in the strict sense of the word at that time.[30] In any case, the country where democracy ranks highest among all the Western nations is, Mūsa goes on to say, Switzerland. No Swiss citizen would understand it if one of their leaders implemented the Nazi or fascist system, because in the Swiss mind all citizens are equal through education and enjoy freedoms guaranteed by the constitution.[31] The occurrence of democracy is, as Mūsā explains, tied to the emergence of a middle class,[32] and "the revolution which we have undertaken in Egypt is the revolution of the middle classes."[33]

Mūsā then refers to Alexander the Great, a popular subject in the Eastern Mediterranean political discourse of the time, because, as an essential figure in the cultural development at the beginning of both the Eastern Roman Empire and the Arab world,[34] he opens a window for utopian fantasies.[35] Yet, Mūsā struggles with the fact that once he had arrived in Egypt in the fourth century BC, Alexander was declared by the local priesthood to be a son of the Godhead Amun. When a king becomes God-like, Mūsā reflects, then the people's belief depends solely on him, so that, when he is toppled, they return to unbelief and atheism.[36] Unbelief and atheism were rejected almost completely in the Middle Eastern political discourse, and religion was not easily abandoned.

An eminent scholar such as Ṭāhā Ḥusayn (1889–1973) has pointed out that Alexander the Great played a salient role not only because he spread the Greek spirit among different cultures, but also because he blended them together into a new civilization that became the foundation of his own empire. Although the

---

28 Ibid., 145.
29 Ibid., 151.
30 Ibid., 154.
31 Ibid., 146.
32 Ibid.
33 Ibid., 150.
34 "Ο βυζαντινός κι ο αραβικός πολιτισμός αποτελούν, κατά την άποψη διάσημων ισλαμολόγων, δύο κλάδους μιας και της αυτής πολιτιστικής κοινότητας, η οποία δημιουργήθηκε μετά την κατάσταση του Αλεξάνδρου από την συνάντηση του Ελληνισμού με τον κόσμο της Ανατολής." Papoulia, *Από την αυτοκρατορία στο εθνηκό κράτος*, op. cit., 40.
35 Victor Tcherikover, *Hellenistic Civilization and the Jews* (Peabody/Mass.: Hendrickson, 1999), 359.
36 Mūsā, *mā hiya an-nahḍa?*, op. cit., 146.

idea of cultural blending was restricted to the Greeks, Macedonians and Persians, it spanned two continents and comprised art, culture and philosophy. This made Alexander much more than an army commander; he was, rather, a pioneer of thought whose impact has never since been emulated in history.[37] In his 1938 book on *The Future of Culture in Egypt*, Ṭāhā Ḥusayn supports the liberation of the individual even in terms of religion. He, too, refers to Alexander the Great, who, according to Ḥusayn, paved the way for reason in every country that he conquered. He praises Alexander for uniting the human mind and nations, as well as for creating a common civilization based on Greek culture. We must therefore conclude that Christianity helped fragment what Alexander had once assembled.[38]

When the pagan tribes, i.e., the Goths and the Sassanides (Persians) attacked Rome, they became absorbed by Christianity, which spread further to the north of the continent and made the continent entirely Christian, before finding a rival in Islam. Both religions resorted to the ancient Greek heritage, which made them trailblazers of contemporary thought.[39] Although Ḥusayn's general attitude towards Islam is apologetic, he takes a more secular stance in his overarching worldview.[40] In his 1921 treatise on the *State of the Athenians* (*niẓām al-Athīniyīn*), he revealed his support for the republic and rejected monarchy as an oppressor of the individual and an entity that did not represent all classes of society.[41]

When Ḥusayn speaks out in favor of using the principles of Aristotelian to create a polity of equality that keeps all extremes in check, then the idea of civil society is not far off. What concerns him most is not general and equal participation in politics, but the constitution of a government that "befits the peo-

---

37 Ṭāhā Ḥusayn, *qādat al-fikr* [Cairo 1925], in idem, *al-majmūʿa al-kāmila li-muʾallafāt al-Duktūr Ṭāhā Ḥusayn*, (Beirut: Dar al-Kitab al-Lubnani, 1973–1983), vol. 8: 175–280, here 266, 268 الاسكندر إذن قائد من قادة الفكر، بل هو زعيم من زعماء قادة الفكر، بل هو اشد قادة الفكر القدماء إنتاجاً واكثرهم نفعاً op. cit., 268. Cf. Michael Kreutz, *Das klassische griechische Schrifttum in der Rezeption der arabischen Nahḍa* (unpublished M.A. thesis, Ruhr-University of Bochum, 1999), 88.

38 Kreutz, *Arabischer Humanismus*, op. cit., 68–70. Ḥusayn, vol. 8, 266–7; cf. Kreutz, *Das klassische griechische Schrifttum*, op. cit., 85. لم يكن يريد ان يفتح الارض وحدها، وإنما كان يريد ان يفتح معها العقل، بل قل إنه اينما كان يفتح الارض تمهيداً لهذا الفتح العقلي Untranslatable pun with *fataḥa* "to conquer" and *fataḥa* "to open." Ḥusayn, vol. 8 267; cf. Kreutz, op. cit., 85 fn. 588.

39 Franz Georg Maier, *Die Verwandlung der Mittelmeerwelt* (Frankfurt/ Main: Fischer [= Fischer Weltgeschichte vol. 9], 1968), 110–1 Ṭāhā Ḥusayn, *niẓām al-Athīniyīn* [Cairo 1921], in idem, *al-majmūʿa al-kāmila li-muʾallafāt al-Duktūr Ṭāhā Ḥusayn* (Beirut: Dar al-Kitab al-Lubnani, 1973–1983), vol. 8: 281–384, here 281.

40 Kreutz, *Arabischer Humanismus*, op. cit., 79, 82.

41 Ḥusayn, *niẓām al-Athīniyīn*, op. cit., 319.

ple" to the effect that "every government, whichever it might be, is good if it serves the people's spirit and its interests."[42] Ḥusayn not only champions democracy as a concept that does not allow discrimination against any belief system, but also stresses that its historical roots lie in the ancient *polis*, which was built around the freedom of the individual.[43] Moreover, he praises Europe both for its technological progress and for following the Cartesian ideal that is based on doubt as the starting-point for all critical thinking, including thinking about religion.[44]

As already mentioned, Ṭāhā Ḥusayn promoted a government that would befit the people[45] so that Aristotle[46] could be a starting-point for a republic. This was because neither the Greeks nor the Romans had experienced despotism during the time of the *polis*, when the political landscape was shaped by small independent entities that only came together briefly in times of danger and that were otherwise reluctant to support any kind of political "strongman."[47] However, this is a very idealistic view, since Athenian democracy did not have a positive notion of the individual, and nor did it hold individual freedom in high esteem.[48]

In any case, Ḥusayn turns out to be a humanist insofar as he repeatedly emphasizes the importance of the individual for the civilized world. He sets forth to explain that the emergence of the individual as a social value can be traced back to the demographic changes in the Eastern Mediterranean during the ninth and eighth centuries BC, when Greek colonies emerged all over Asia, Italy and Sicily and their settlers came into contact with other peoples and cultures. This resulted in the rise of strong individuals who refused to bow to the government, while their resistance was shored up by rationalist currents legitimizing individualism. Ḥusayn makes clear that this development was successful against all the odds, and that the Arab world can also succeed in its own individual liberation, since its cultural resources include elements of classical Greek thinking.[49]

42 Ibid., 359.
43 Ḥusayn, *qādat al-fikr*, op. cit., 207.
44 Kreutz, *Zwischen Religion und Politik*, op. cit., 121–2.
45 فكل الحكومة مهما تكن صورتها، خير إذا لادمت روح الشعب ومنافعه Ḥusayn, vol. 8, 359; cf. Kreutz, *Das klassische griechische Schrifttum*, op. cit., 82 fn. 564.
46 كان لها الأثر الأكبر في تكوين العقل العربي الإسلامي ... فلسفة ارسطاطاليس Ḥusayn, *qādat al-fikr*, op. cit., 260; cf. Kreutz, *Das klassische griechische Schrifttum*, op. cit., 78 fn. 540.
47 Ḥusayn, *qādat al-fikr*, op. cit., 278; cf. Kreutz, *Das klassische griechische Schrifttum*, op. cit., 87–8.
48 Norbert Wokart *Kontaminationen: Antike Spuren in unserem Denken* (Würzburg: Königshausen und Neumann, 2014), 126.
49 Ḥusayn, *qādat al-fikr*, op. cit., 203–204; cf. Kreutz, *Arabischer Humanismus*, op. cit., 55–6.

It is little wonder that contemporary Greek thinkers of the time should emphasize the historical continuity that their own culture had with the culture of the antiquity. This must also be seen against the backdrop of the denial made in 1830 by the German historian Jakob Philipp Fallmerayer that there was any such historical continuity, and of his claim that the Greeks of his time were merely a superficially Hellenized people whose roots lay in the Slavonic substrate of the Balkans. Fallmerayer's views caused an intellectual earthquake among the Greek elite of the time.[50] The poet and nationalist historiographer Spyridion Zambelios (1815–1881), who was educated in Germany before going to Italy and Switzerland, was the first to proclaim a historical continuity leading from the antiquity to the present, which he called "Greek-Christian" (ἑλληνοχριστιανικό).[51]

This controversy found a late echo in a 1914 article in the Egyptian journal *al-Hilāl.*, It refers in its report on the Greeks to a group of intellectuals who still believed that the ancient Greek people and the ancient Greek language had died out, and that there was therefore a huge difference compared to contemporary Greeks and their language. Yet, the article goes on to argue that all contemporary studies dispute this point of view, and, in contrast, see many connections between the Greeks of the past and the Greeks of the present, as well as between their languages. Although there are differences, we cannot speak of two distinct entities.[52] It seems reasonable for the reader to relate this account to Arabic culture and language, which should be understood with the fact in mind that the contemporary vernacular and the classical tongue are so distinct from each other that a theory had emerged by the end of the nineteenth century that suggested that the contemporary language had in fact derived from the ancient Carthagian language.[53]

The journal also discusses the historical relationship between East and West, pointing to the split between the Eastern and the Western church in 1053, which marked the origin of the quarrel between East and West (*manshaʾ an-nizāʿ bayna ash-sharq wa-l-gharb*).[54] Contrary to the standard narrative that

50 Vakalopoulos, *Griechische Geschichte*, op. cit., 160–1. Ioannis Zelepos, *Die Ethnisierung griechischer Identität, 1870–1912: Staat und private Akteure vor dem Hintergrund der 'Megali Idea'* (Munich: R. Oldenbourg, 2002), 78.

51 Evi Petropoulou, *Geschichte der neugriechischen Literatur* (Frankfurt/Main: Suhrkamp, 2001), 56, 64.

52 N.N., "al-Yūnān," op. cit., 269.

53 Cf. Kreutz, *Das Ende des levantinischen Zeitalters*, op. cit., 144–5.

54 N.N., "al-Yūnān," op. cit., 275–6. Similarly in *al-Hilāl*, XXIV, No. 4, Jan. 1916: 268–77, here 275–6.

the end of crusades was associated with Saladdin, a 1916 article in the same journal points to the fact that it was the return under Michalis Palaiologos of the Eastern Roman state in 1261 that caused the crusader state to begin to dwindle and then finally to disappear. The Palaiologos family remained in power until Constantinople was defeated in 1453.[55] This must have been a fresh perspective for the Arab reader, shedding a different light as it did on the Muslim conquest of Constantinople. By extension, the loss of Constantinople must be seen as representing the loss of a bulwark against the crusaders.

Generally, the journal showed a lot of support for the Greeks. When in 1770 the Russian empress Catherine the Great encouraged the Greeks to pursue their own struggle to liberate themselves from the Turks, she did not conceal the religious bonds that existed between Greece and Russia. The reader is told that the Greeks failed and were punished, but never gave up on their cause. The French Revolution is depicted as the most important factor in the quest for independence. Since then, Europeans, such as the philosopher Voltaire and the poet Lord Byron, had begun to show an interest in the Greek cause and to demand the liberation of the Greeks. The article then gives an account of the Greek uprising in 1821, the help provided by Russia, and the leadership of Alexandros Ypsilantis. The revolution had spread like wildfire to begin with, but was quenched by the sultan and his aide, Muḥammad ʿAlī, who achieved victory in the battle of Messolongi, when finally the European states came to the rescue of the Greeks. Two years later, the same journal dedicated a whole volume to the phenomenon of European philhellenism, including Byron, who is portrayed as having loved the Greeks so much that he had fought with them until his own death in 1824.[56]

The Greek nationalist Ionos Dragoumis compared Hellenism (i.e., Greek culture) to "a big and fervent river," one whose water may change over time but always flows with the same element. The river is Hellenism and its waters are "the Greeks who are passing by and who are running one after the other, although they are not the same; it is not the same Greeks who are running by but the river consists of the same element, i.e., Greeks. It is always Greeks flowing the river, which is called Hellenism, and they are flowing and flowing because they don't cease to exist. Therefore Hellenism lives on and does not vanish."[57]

---

55 N.N., "al-Yūnān," op. cit., 276.

56 Ibid., 276.

57 "Σάν ποταμός μεγάλος καί ζεστός τρέχει ὁ Ἑλληνισμός μέ τούς ἄρχοντές του. [...] Τό νερό δέν εἶναι τό ἴδιο, μά ἔχει τά ἴδια συστατικά (H2O). Ἔτσι καί τό νερό τοῦ Ἑλληνισμοῦ, τό νερό τοῦ ποταμοῦ πού λέγεται Ἑλληνισμός, τό νερό αὐτό που εἶναι οἱ Ἑλληνες που περνοῦν καί τρέχουν ὁ ἕνας ὕστερα ἀπό τόν ἄλλο, δέν εἶναι πάντα τό ἴδιο· δέν τρέχουν οἱ ἴδιοι Ἕλληνες, ἀλλά ἔχει τά ἴδια συστατικά, ἀλλά τρέχουν Ἕλληνες. Ἕλληνες πάντα τρέχουν μέσα στό ποτάμι, που λέγεται

Dragoumis emphasizes the fact that the Greeks are native, whether they were "colonists, settlers, immigrants, emigrants." It is for this reason that the Greeks belong to Rumelia, while "in America they are foreigners."[58]

Connecting the dots of the past, Dragoumis also speaks of contemporary "Byzantine" cities, where he used "to stroll around the old streets and enter old homes." He describes with anger and with an overtly racist attitude the town of Philippopolis, which is the present Bulgarian town of Plovdiv, as being "entered by barbarians of all kinds, by Bulgarians, Jews, Armenians and Gypsies." The term "Byzantines" clearly has a negative connotation when he speaks of "enslaved Greeks" who are not real Greeks but Byzantines (whom he feels like a stranger among).[59] So, for Dragoumis, the Eastern Roman Empire represents a kind of lower Greekness, a Greekness of the past.

Dragoumis is for the same reason less interested in the *katharevousa*, the "pure" version of Greek, which he classifies as "a remnant of Byzantine life and spirit." He is not completely opposed to this mode of Greek because he feels "some Byzantine spirit" within himself, but he prefers the *dimotiki*, the vernacular Greek, which is said to be "more vivid, the language of the future."[60] He also speaks of a third version of Greek, which he calls *oloumeni* ("colloquial"), but he does not explain how it differs from the *dimotiki*.[61] He then goes on to argue that the *katharevousa* is the language of public service while the *dimotiki* is the language of life – although the language of life is also the language of the slave,[62] since it was spoken under Turkish rule. However, the *dimotiki* is also the

---

Ἑλληνισμός, καί περνοῦν καί περνοῦν, γιατί εἶναι περαστικοί, μά ὁ Ἑλληνισμός μένει καί δέν περνᾶ." Dragoumis, Ὁ Ἑλληνισμός μου καί οἱ Ἕλληνες, op. cit., chap. A, § 10: 25.

58 "Ἄλλες χῶρες εἶναι ἐκεῖνες, ὅπου οἱ Ἕλληνες εἶναι ἰθαγενεῖς, καί ἄλλες ἐκεῖνες, ὅπου εἶναι ἄποικοι, πάροικοι, μέτοικοι, μετανάσται, ἀπόδημοι. Στήν Ἀνατολική Ρωμυλία οἱ Ἕλληνες εἶναι ἰθαγενεῖς. Στήν Ἀμερική εἶναι πάροικοι." Ibid., chap. B, § 2: 38.

59 "Μ' ἀρέσει ἐγώ, ὁ νέος Ἕλλην, νά πηγαίνω στίς βυζαντινές πολιτεῖες καί νά περιπατώ στούς παλιούς τούς δρόμους καί νά μπαίνω στά παλιά τά σπίτια [...]. Στή Φιλιππούπολη μπῆκαν οἱ βάρβαροι, ὅλων τῶν εἰδῶν, Βούλγαροι, Ἑβραῖοι, Ἀρμένηδες καί Γύφτοι. [...] Περίεργο με φαίνεται ἐγώ, ὁ νέος Ἕλλην, νά σχετίζωμαι με Βυζαντινούς. Οἱ περισσότεροι ὑπόδουλοι Ἕλληνες δέν ἔγιναν ἀκόμη Ἕλληνες καί εἶναι Βυζαντινοί. Μέ φαίνεται πως εἶμαι ξένος ἀνάμεσά τους." Ibid., chap. B, § 3: 38.

60 "Καί ἡ γλῶσσα ἡ καθαρεύουσα εἶναι ἀπομεινάρι ζωῆς καί πνεύματος βυζαντινοῦ. Τήν ἐννοῶ, γιατί κ' ἐγώ ἔχω ἀρκετό βυζαντινό πνεῦμα, μά ἐννοῶ καί τή δημοτική που εἶναι πιό ζωντανή, ἡ γλῶσσα τοῦ μέλλοντος. [...]" Ibid., chap. B, § 10: 44.

61 "Τρεῖς γλῶσσες εἶναι τώρα στό Ἑλληνικό που "ἔχουν χάριν", ἡ καθαρεύουσα [...], ἡ ὁμιλούμενη [...], καί ἡ δημοτική [...]." Ibid., chap. Γ, α Athens 1905, § 11: 53.

62 "Ὁ Γευτός ἐδῶ μέ εἶπε: "μίλησα στή γλῶσσα τῆς ὑπηρεσίας (καθαρεύουσα)". Καί ἔτσι εἶναι. Ἡ γλῶσσα ὅμως τῆς ζωῆς ποιά εἶναι; [...] Λοιπόν ἡ γλῶσσα τοῦ δούλου δέν εἶναι ἡ γλῶσσα τῆς

language of the future for it re-connects the Greeks with their ancient past. The fact that the *dimotiki* is spoken but is not "pure" is not a problem because "the written languages in all parts of the world are somewhat artificial."[63] He describes his and the dimoticist's *dimotiki* as a language "that would have been if there had been no teachers and no spirit of literacy."[64]

Still, Dragoumis himself was surprised to learn that the Greek of his time had to some extent served as a *lingua franca:* "The Greek language, language of the salons like French in Greece and in Europe, is in the mouth of all Frank-olevantines, Armenians and Jews," he argues, adding that he frequented a salon in Dedeağaç (now Alexandroupoli in Greece), where he found "Italofranks, Gallofranks, Armenocatholics, Greeks, Jews and their wives," who all spoke Greek.[65] That said, he concludes that "the language question is not only linguistic but social. The growing number of dimoticists among the nation means that the Greek mind is slowly changing and is beginning to feel things which it did not feel before." He lashes out against the Philhellenes and the educated Greeks, who adhered to a lie by telling the Greek population that they were "Pericles, Themistocles, Socrates, and Sophocles," therefore making them "patsies with helmets, spears, sandals and tunics – all spoof, paper and wood, glitter, inlaid with tinsel." When he finally calls the demotic language "an illness," he shows how much he is torn between a distant past "and an even more distant present."[66]

The first writer that actively promoted language reform was sixteenth-century Nikolaos Sofianos of Corfu, who drafted a grammar and translated ancient Greek texts into contemporary Greek, such as Plutarch's "On the education of

ὑπηρεσίας, καί ἡ γλῶσσα τῆς ὑπηρεσίας δέν εἶναι ἡ γλῶσσα τῆς ζωῆς καί ἡ γλῶσσα τῆς ζωῆς εἶναι ἡ γλῶσσα τοῦ δούλου." Ibid., Κραβασαράς Λοκρίδος, 21 May 1904, 68–9.

63 "Ἡ γραμμένη γλῶσσα, σ' ὅλα τά μέρη τοῦ κόσμου, ἔχει κάτι τί τό ἐπιτηδευμένο (artificiel). [...]" Ibid., Κραβασαράς Λοκρίδος, 21 May 1904, 73.

64 "Ἐμεῖς οἱ δημοτικιστές γράφουμε τήν ἑλληνική γλῶσσα ὅπως θά ἦταν, ἄν δέν ὑπάρχανε δάσκαλοι καί δασκαλικό γραμματισμένων πνεῦμα. [...]" Ibid., Ἀθῆναι 1905, 129.

65 "Ἡ ἑλληνική γλῶσσα, γλῶσσα τῶν σαλονιῶν ὅπως τά γαλλικά στήν Ἑλλάδα καί στήν Εὐρώπη, εἶναι στά στόματα ὅλων τῶν Φραγκολεβαντίνων, τῶν Ἀρμενίων καί τῶν Ἑβραίων. Ἤμουν χθές σ' ἕνα σαλόνι ἐδῶ, στό Δεδέαγατς· ἦταν ἐκεῖ Ἰταλόφραγκοι, Γαλλόφραγκοι, Ἀρμενοκαθόλικοι, Ρωμιοί, Ἑβραῖοι καί οἱ γυναῖκες τους. Ὅλοι μιλοῦσαν ἑλληνικά." Ibid., Δεδέαγατς [= Alexandroupoli – Dedeağaç] 1905, 86.

66 "Τό ζήτημα τό γλωσσικό δέν εἶναι μόνο γλωσσικό εἶναι κοινωνικό. Μέ τό νά πληθαίνουν οἱ δημοτικιστές στό ἔθνος σημαίνει πως ἀλλάζει σιγά σιγά τό μυαλό τῶν Ἑλλήνων καί ἀρχίζει καί νοιώθει πράματα που πρίν δέν τά ἔνοιωθε. [...] Οἱ φιλέλληνες καί οἱ μορφωμένοι τοῦ γένους μᾶς εἶπαν ψέματα, μᾶς εἶπαν πως εἴμαστε Περικλῆδες, Θεμιστοκλῆδες, Σωκράτηδες, Ἀριστοτέληδες καί Σοφοκλῆδες. Μᾶς μασκάρεψαν μέ περικεφαλαῖες, μέ κράνη, μέ δόρατα, μέ σάνδαλα, μέ χιτῶνες – ὅλα ψεύτικα, χάρτινα καί ξύλινα, χρυσωμένα, μέ κολλημένα χρυσόχαρτα. [...] Ἡ δημοτική γλῶσσα εἶναι μιά ἀρρώστια [...]" Ibid., Ἀθῆναι 1905, 123, 125.

children" (Περὶ παίδων ἀγωγῆς),[67] which was published in 1544. The sixteenth-century language became a common medium for texts related to the Christian faith, these texts becoming part of an emerging religious humanism driven by writers such as Maximos Margounios (1530–1602) and Meletios Pegas (1535–1602). A printing press at the patriarchate of Constantinople was established by patriarch Kyrillos Loukaris (1572–1638), who in 1638 also encouraged the translation of the Gospels into contemporary Greek.[68]

The following century saw the publication of a considerable number of books in contemporary Greek.[69] The Greek language had undergone a number of major transformations, and the gap between the language of the New Testament and the vernacular of the nineteenth century was quite significant. A first effort to translate the Gospels "from the Hellenic into the Roman language" (άπό ἑλληνικήν γλῶτταν εἰς ρωμαίικην) was made by Maximos Kallioupolitis in 1632.[70] Thus, "Hellenic," which in Eastern Roman times was used to denote ancient Greek, was now used for the Greek of the New Testament, while "Roman" (Romaic) became the term for the contemporary vernacular language. Another translation of the New Testament into the contemporary language was carried out by Maximos of Gallipoli under the auspices of the aforementioned patriarch Kyrillos Loukaris. His translation was printed in Leiden (Netherlands) in 1683.[71]

Meanwhile, Eugenios Voulgaris used to write in a scholarly Greek indebted to the ancient Attic standard,[72] since the idea that philosophical issues could be discussed in the vernacular was alien to him. But the Greek independence movement regarded the translation of literature as an important tool for the cultural revival of the Greek nation.[73] Under the auspices of the Phanariot Nikolaos Mavrokordatos (1680–1730), who had gathered German and Greek scholars around

67 The Περὶ παίδων ἀγωγῆς is the first chapter of the first book of Plutarch's *Moralia*.
68 Strakhov, *The Byzantine Culture*, op. cit., 59–60. There is little known of the Greek printing in Constantinople in the 18th century. Clogg, "The ‚Dhidhaskalia Patriki'," op. cit., 101.
69 Strakhov, *The Byzantine Culture*, op. cit., 60–4.
70 Hassiotis, "From the ‚Refledging' to the ‚Illumination'", op. cit., 53–4. "Romaiki" was in use at least until the early 20th century, v. Odysseus, *Turkey in Europe* (London: Edward Arnold, 1900), 299. As soon as in 1547 Jews in Constantinople translated the Hebrew Bible into contemporary Greek, v.. George N. Hatzidakis, *Einleitung in die neugriechische Grammatik* (Leipzig 1892, repr. Hildesheim and New York: Olms, 1977), 244.
71 Heyer, *Die Orientalische Frage*, op. cit., 2; cf. Julius Wiggers, *Geschichte der Evangelischen Mission* (Hamburg and Gotha: Perthes, 1845), vol.1 (2), 233.
72 Henrich, "Als Denker Archaist, als Dichter auch Demotizist," op. cit., 99.
73 Ἐλσυρικτέον ἄρα τὰ χυδαϊστὶ φιλοσοφεῖν ἐπαγγελόμενα βιβλιδάρια – "Auszuzischen folglich sind die armseligen, Philosophie zu behandeln vorgebenden Büchlein." Henrich, "Als Denker Archaist, als Dichter auch Demotizist," op. cit., 102.

him, it was Dimitrios Prokopios (Pamperis Moschopolitis, 17th/ 18th century) who composed the first history of modern Greek literature.[74]

However, the translation of the Bible remained at the center of language reform. This was not initiated until the early nineteenth century, when Adamantios Korais turned to the British and Foreign Bible Society (BFBS) for this purpose A first attempt to modernize the New Testament had already been made in 1703, but Korais felt the need for an updated version purified of Turkish loanwords. It seems that he was not aware that such an improved version already existed, funded by the Prussian queen Sophia-Louise. In any case, it was the Russian Bible Society that succeeded in this effort and published a version of the New Testament in vernacular Greek that was distributed among the Greek colonies in southern Russia and the Ottoman Empire.[75]

The Bible in its modernized version then became among the Christian population a catalyst for undermining Ottoman rule. The Christian population were strengthened by the American missionary headquarters in the Balkans, which preferred any other Christian rule in that area to Ottoman suzerainty. In a twist of irony, the missionaries were not welcome among the Greek and Serbian communities, but were met with anger, and their schools and churches reportedly burned down by Serbian fanatics.[76] In any case, the translations could not fall on fertile ground among the masses since the level of education and literacy in the Balkans was low. The Habsburg dynasty aimed at increasing the number of schools in Bosnia and the region of Herzegovina, but that attempt failed when the peasantry proved reluctant to send their children to school.[77] Yet, the missionaries did not cease their eager dissemination of copies of the Bible, and, in doing so, made the printing press popular. Their contribution to the reform

---

74 Von Maurer, *Das griechische Volk*, Erster Band [1st vol.], op. cit., 23–4. Paschalis M. Kitromilides, *Enlightenment and Revolution: The Making of Modern Greece* (Cambridge/Mass. and London: Harvard University Press, 2013). 31.

75 Heyer, *Die Orientalische Frage*, op. cit., 1–5. Nikifor Theotoki cooperated with Pinkerton in founding a central Bible Society for the Ionian islands, v. ibid., 5; William Jowett, *Christian Researches in the Mediterranean from 1815 to 1820: In Furtherance of the Objects of the Church Missionary Society*, (London: Seeley and Hatchard, 1822), 43.

76 Ömer Turan, "American Protestant Missionaries and Monastir, 1912–17: Secondary Actors in the Construction of Balkan Identities," in *Middle Eastern Studies*, Vol. 36, No. 4, Oct. 2000: 119–36, here 120–34.

77 Clark, *Sleepwalkers*, op. cit., 75–6.

of the Greek language found a strong supporter in Patriarch Gregory V, who was also open to translations of English texts into Greek.[78]

The Archimandrite Hilarion, head of the Patriarchal press in Constantinople, published a translation of his own. After initially being given the green light by the archbishops of the Sinai and Thessalonika, as well as by the Holy Synod, it failed in a second review by the British and Foreign Bible Society (BFBS), which described it as being long-winded and lacking concision. When Hilarion learned that his translation had been altered without his permission, he intervened in the printing process.[79] The text finally reached the printing press, beginning in 1827 with the Gospels, but other bishops, having read the translation for the first time, expressed additional criticism. Translations of the scripture into contemporary Greek were generally met with suspicion. Hilarion rejected the objections and justified his method by pointing to the church's missionary tradition of bringing the holy scriptures to the masses, thereby making translations indispensable.[80]

A similar conflict broke out when it came to the translation of the Hebrew Bible. The British and Foreign Bible Society (BFBS), which was located on Corfu, rejected a first attempt by Hilarion that was based on the Septuagint, so that he suggested translating directly from the Hebrew. A group of translators tackled the enterprise, which was not without controversy, since not everyone in the church liked the idea of privileging the Hebrew Bible over the Greek Septuagint. The translation was finished in 1844.[81]

Eastern European languages were already covered, partly due to the Russian Bible Society, which had published a Romanian and a Bulgarian version in the

78 Heyer, *Die Orientalische Frage,* op. cit., 7. Gregorios was in contact with John Pinkerton, v. ibid., 6; cf. Wiggers, *Geschichte der Evangelischen Mission,* op. cit., vol. 1 (2), 233. On Pinkerton's role v. Jowett, *Christian Researches in the Mediterranean,* op. cit., 363–5.

79 "Our object has been to cut off excrescences, and to make the translation as terse and as close to the original as possible." Rev. H. D. Leeve, Feb. 8, 1821, in *Report of the British and Foreign Bible Society,* vol. 17 (1821): 66. Their names were G.C. Renouard and T.P. Plat, v. Heyer, *Die Orientalische Frage,* op. cit., 7–8.

80 Heyer, *Die Orientalische Frage,* op. cit., 8; cf. the letter by Hilarion to the Patriarch of Constantinople, Anthimos, and the Synod of the Greek Church, publ. in Engl. translation in *The Twenty-First Report of the British and Foreign Bible Society* (London: Augustus Applegath, 1825), 154.

81 Samuel Bagster, *The Bible of Every Land: A History of the Sacred Scriptures in Every Language and Dialect Into Which Translations Have Been Made* (London: Sam. Bagster, 1848), 203; Heyer, *Die Orientalische Frage,* op. cit., 8–10. One of the translators, Neophytos Vamvas, was in contact with Adamantios Korais, v. ibid.; cf. Hilarion's letter (v. footnote 80), in *The Twenty-First Report of the British and Foreign Bible Society,* op. cit.

early decades of the century, partially funded by the BFBS. The American Bible Society, founded in 1816, published another translation of the Gospels by Elias Riggs in 1855. Translations of the Bible into contemporary Greek became an important means of support for the Greek national movement, since such translations vindicated their call for a general reform of language.[82] The translations were joined by translations of secular literature, so that, between 1850 and 1880, there was about one translation per decade.[83] German often served as an intermediary language when French, English or even Russian literature was translated into local languages.[84] There was also some interest in German literature itself, so that, in the mid-nineteenth century, there were several translations of Goethe's *Werther* into demotic Greek. This book is of special significance since it paved the way for the influx of Ossian into Greece, inspired by *The Songs of Selma* included in *Werther.*[85]

As mentioned, the Protestant mission found it difficult in the Christian Orthodox world, and it did so also among Catholics. It was the Maronite Patriarchate on the southern shores of the Mediterranean that in 1824 condemned both the *King James Bible* and the Protestant mission for being heretical.[86] However, the Bible translation movement remained unimpressed and initiated translations into Albanian and Turkish.[87] Being the language of the Quran, Arabic now became the language of the Bible, too.[88] As Barqāwī (1988) has pointed out, the transmission of cultures is a universal phenomenon, which means that it is difficult to find a people who formed their culture in isolation from other cultures.

---

82 Heyer, *Die Orientalische Frage*, op. cit., 11–2. In Constantinople in 1854 an corresponding agency called "Bible House" was established, v. ibid., 12.

83 Ioanna Petropoulou, "From West to East: The Translation Bridge. An Approach from a Western Perspective," in *Ways to Modernity in Greece and Turkey*, ed. Anna Frangoudaki and Çağlar Keyder (London: IB Tauris, 2007): 91–112, here 98.

84 Matl, op. cit., 441; cf. Kreutz, *Das Ende des levantinischen Zeitalters*, op. cit., fn. 194.

85 Kathleen Ann O'Donnell, "How Twentieth Century Greek Scholars Influenced the Works of Nineteenth Century Greek Translators of ‚The Poems of Ossian' by James Macpherson," in *Athens Journal of Philology* Vol. 1, No. 4 (December 2014): 273–84, here 277.

86 A. L. Tibawi, "The Genesis and Early History of the Syrian Protestant College," in *American University of Beirut Festival Book* (Festschrift), ed. Fuad Sarruf and Suha Tamim (Beirut: Eastern Printing, 1967): 257–94, here 261–4.

87 Rev. H. D. Leeve, Feb. 8, 1821, in *Report of the British and Foreign Bible Society*, Band 17 (1821), 66.

88 Barqāwī, *muḥāwala*, op. cit., 87.

Yet, the great impact that Western civilization had on the Arabic Nahḍa has a specific nature.[89]

Salāma Mūsā reminds his readers of the old wisdom that words are the seeds of thoughts, adding that words are also the seeds of deeds. Therefore, the terms 'freedom', 'equality', and 'fraternity', which the French *philosophes* of the eighteenth century used in their writings, were seeds of both thoughts and deeds. Given that "language is the basis of culture," it becomes evident that first language must undergo a process of modernization before society is able to, which means that old terms must be assigned to new meanings, or that new terms need to be coined. Mūsā points to Europe, where the advancement of the vernacular languages was the outcome of the medieval use of Latin, which no longer seemed appropriate. We can find something similar, he says, in China.[90] The message is clear: if the Arabs wish to participate in modernity, then they have to modernize their language, too.

The growing interest of Arab intellectuals in the ancient pre-Islamic heritage also included epic poetry. Ismāʿīl Adham, together with the Lebanese politician and intellectual humanist Felix Fares (Fīlīks Fāris, 1882–1939), who had earned some recognition for his translation of Nietzsche's *Thus spoke Zarathustra*,[91] expresses his surprise at the fact that the Arabs who had translated a large part of Greek literature showed a complete lack of interest when it came to Greek poetry and the fine arts. Fares explains this by pointing not to the Quran but to the Arab's different mentality: while Greek culture is based on a sense of the chasms of life, the Arab culture is less so which is the reason that there is no Pindar, Ovid or Homer in Arabic literature.[92] Adham also agrees with Fares that the non-exis-

89 فالمثقف اليوناني القديم وجد في علم أساطير الشرق .إننا في واقع الحال لسنا أمام ظاهرة استثنائية في التاريخ، تاريخ انتقال الأفكار وغدت الفلسفة .وغدت بعد هضمها جزء اً لا يتجزأ من العلم والفلسفة الأغريقيين .مادة ستستجيب لحاجات المجتمع اليوناني الديمقراطي وفي هذه الفلسفة وجد عصر النهضة سلاحاً لمواجهة .اليونانية بدورها سلاحاً ايديولوجياً مهماً في الفلسفة العقلانية العربية الإسلامية ولكن حسبنا القول إن ظاهرة انتقال الثقافات .. إن تفسير هذه الظواهر يحتاج ولا شك إلى وقفة خاصة .الكنيسة في بعض الحالات ولكن .هي من العمومية بحيث يصعب أن نجد شعباً صاغ ثقافته بمعزل عن التأثير الذي مارسته ثقافات شعوبٍ أخرى عليه الظاهرة التي نحن بصدد تناولها ـ أي ظاهرة الأثر الكبير الذي مارسته الايديولوجيات والثقافة الغربية بعامة على إيديولوجيات وثقافة عصر النهضة العربية تمتلك خصوصيتها Ibid., 43.

90 Mūsā, *mā hiya an-nahḍa?*, op. cit., 142.

91 Salim Mujais, *Antoun Saadeh: A Biography*, Vol. 1: *The Youth Years* (Beirut: Kutub, 2004), 92. On Fares cf. Keith D. Watenpaugh, *Being Modern in the Middle East: Revolution, Nationalism, Colonialism, and the Arab Middle Class* (Princeton/New Jersey: Princeton University Press, 2006), 113.

92 ومما يجدُر ذكرُه هو أنّ العرب حين اقتسبوا من تراث اليونان ما يعزِّزون به تفكيرهم العلمي لم تستهويهم الثقافة اليونانية ولا حضراتهم الادبية، إذ احسوا بما بين الحضارة اللتي كانت تتمخض في شعورهم وتقديرهم للحياة وبين حضارة اليونان الإجتماعية من مهاو سحيقة، فأعرضوا عن شعرهم وموسيقاهم ونظم اجتماعهم، لذلك لا تجد في شعر العرب شيئاً من إلهام بيندر واوريبيد وحوميروس Adham, "bayna al-gharb wa-sh-sharq," op. cit., 143. Fares' interest in the ancient Greek literature and culture is somewhat typical of his time a is his interest in the question

tence of epics, drama and visual poetry in Arabic culture is due to its "subjective" (*dhātīya*) nature, which runs contrary to the Greek "objective" (*mawḍū'īya*) culture. Adham and Fares believe that Arabic culture cannot shift to objectivity because its mentality lacks the resources to perform such a (necessary) task.[93] We would nowadays dismiss such criticism as untenable, pre-scientific and culturalist. Yet, we should recognize that such ideas and interests as expressed by Adham and Fares, when seen against the background of their time, do reflect a critical stance towards their own culture, a stance driven by a strong sense of modernization and reform.

One such example is the contemporary Egyptian author 'Alā' al-Aswānī, who states, in direct opposition to Adham, that Islam calls for the application of objective scientific methodology in understanding the world.[94] According to Zaqzūq (1984), a renowned contemporary Egyptian theologian (v. p. 29), it is the highest goal of Arab philosophers to "acquaint the world with a holistic theory of the unity of being which satisfies both religion and reason." Zaqzūq also emphasizes that contemporary Muslim scholars have highlighted the significance of Islam for the advancement of human thought in general and its own adoption of Persian and Eastern Roman cultural elements, i.e., Sufism and nationalism. This leads him to conclude that East and West have met in one melting pot – yet another holistic theory of the unity of being.[95]

It is important to note in this context that translations of Homer's epic poetry did indeed gain some popularity in the mercantile circles of the Eastern Mediterranean. For Ionos Dragoumis, a common denominator of the Greeks as a "commercial people" is the fact that they are active in writing sonnets and short sto-

---

why epic and drama never really took foot in the Arab literary history. On a discussion among Arab intellectuals of that time v.. Kreutz, *Arabischer Humanismus*, op. cit., here chaps. "Gibt es eine arabische Epik?," 36–41, and "Ödipus und die menschliche Seele," 100–7; also idem, "The Greek Classics in Modern Middle Eastern Thought," in *Judaism, Christianity, and Islam in the Course of History: Exchange and Conflicts*, ed. Lothar Gall and Dietmar Willoweit (Munich: R. Oldenbourg, 2011): 77–92, *passim*. Tawfīq al-Ḥakīm assumed that certrain texts have been spared out because the translators were Syrian monks with no interest in poetry. Tawīq al-Ḥakīm, *al-malik Ūdīb* [Cairo 1949], preface, p. 17, quoted after Kreutz, *Arabischer Humanismus*, op. cit., 101.

93 ولهذا لن تجد في الأدب العربي شعراً قصصياً ولا شعراً تمثيلياً ولا شعراً تصويرياً، لأنّ القصص والتمثيل والتصوير يستلزم الأنسحاب من آفاق الذات إلى رحاب الموضوعية، وليس هذا في مُكنة الذهنية العربية Adham, "bayna al-gharb wa-sh-sharq," op. cit., 143.

94 'Alā' al-Aswānī, "hal taṣluḥ ad-dīmuqrāṭīya li-ḥukm al-muslimīn?," in idem, *maqālāt 'Alā' al-Aswānī*, vol. 2: *hal nastaḥaqq ad-dīmuqrāṭīya?*, (Cairo: Dar ash-Shuruq, 2010): 134–7, here 136.

95 Zaqzūq, op. cit., 24–5.

ries – and, yes, in translating the *Iliad*.[96] Such reflections on the pagan cultural heritage go hand in hand with concepts of identity. Sulaymān al-Bustānī, who himself translated the *Iliad* into Arabic, has pointed out that the Turkish (Ottoman) language is the "language of the rulers," but at the same time is unknown even in parts of the government. Every time that rulers such as sultan Salim I tried to make Arabic the official language, they failed, with every *umma* adhering to its own language – but, where there is no mutual understanding and recognition of cultural differences, there is no melting and mixing.[97] Therefore, the Arabs are likely to see their future as lying in independence from the Ottoman bond.

3.2 Mapping Out New Orders

Sulaymān al-Bustānī was convinced that a people such as the Arabs, who for several centuries had been ruled by the Ottomans, would not be able to establish freedom overnight. Instead, doing so would strengthen the state of anarchy, and the process of self-liberation would end in tyranny by certain groups, which is tyranny much worse than tyranny by an individual. In contrast, he argues, the Ottomans were never enslaved, but instead made conquerors, so that, in an Ottoman context, the idea of freedom that was safeguarded in a constitution could work.[98] Bustānī, who was an ardent supporter of constitutionalism, argues that a deliberative assembly (*majlis shūrawī*) similar to the one defined by the

96 "Ένας ξένος πού μάς κοιτάζει θἄλεγε πως οἱ Ἕλληνες εἶναι "ἐμπορικός λαός". Ὅμως αὐτοί οἱ ἔμποροι εἶναι ἄνθρωποι, καί αὐτοί οἱ ἔμποροι κάνουν καί σονέτα καί γράφουν διηγήματα καί μεταφράζουν τήν Ἰλιάδα, καί καταγίνονται μ' ὅλα τά ζητήματα τῆς ἰδέας [...]." Dragoumis, Ὁ Ἑλληνισμός μου καί οἱ Ἕλληνες, op. cit., chap. B, § 5: 41.

97 وحيث لا يحصل التفاهم لا يحكوم الاندماج والتزامج Bustānī, *ʿibra wa-dhikrā*, op. cit., 152–3. On Bustānī's projekt of translating the *Iliad* cf. Michael Kreutz, "Sulaymān al-Bustānīs *Arabische Ilias:* Ein Beispiel für arabischen Philhellenismus im ausgehenden Osmanischen Reich," in *Die Welt des Islams* vol. 44, 2 (2004): 155–94, *passim*.

98 يقول ارباب السياسة لا يسوغ اطلاق الحرية دفعة واحدة لامة طال عليها عهد الاستعباد، لئلا تستحكم الفوضى، وينتهي الامر ولكن هذا القول، مع ما فيه من الصواب، لا ينطبق على الامة .باستبداد الجماعات، وهو أشد بلاء من استبداد الرجل الفرد بل كانت منذ تألفت تحت لواء السلطان عثمان الغازي، امما فاتحة تحت .العثمانية، فانها ليست بالامة التي رسفت دهرا بقيد الرق وان جميع العناصر التي انضمت تحت لوائها كانت من ذوات الماضي .زعامة العنصر التركي وشعوبا مكافحة ذودا عن حياضها وهذا السلطان محمد الفاتح، مع ما يعزى اليه من القسوة، قد خول رعاياه .وان كثيرين من سلاطينها كانوا ذوي برعايتهم .المجيد المسيحيين الاسرائليين من حرية الدين والتصرف بالاحوال الشخصية Bustānī, *ʿibra wa-dhikrā*, op. cit., 90. On freedom and constitution, Midḥat, Ḥusayn ʿAwnī and Rushdī, cf. ibid., 91–2.

sharīʿa had already been in place in ancient Greece.[99] He goes on to argue that, when the Ottoman constitution was proclaimed in 1876, it was the Shaykh al-Islām who legitimized the constitution by declaring that its ideal be engraved on the heart not only of every shaykh and priest, but also of every Muslim and Christian, and of every Ottoman and human.[100] According to the Shaykh al-Islām, the constitution was in line with the Quran and there was nobody to deny that it was also in line with the Torah and the New Testament.[101] Bustānī is full of praise for the constitution. When it was declared, he writes, the imam, the priest and the rabbi hugged each other, while the world witnessed their fraternity and their eyes were awash with tears of joy.[102]

The constitution put an end – at least in theory – to the old system of sectarianism, where every religious community had sought the protection of a great power. In Damascus, Western consulates which had existed since the 1830s provided Christians and Jews with additional protection in terms of commerce, which is the reason that they enjoyed reduced import tariffs.[103] Relationships between the religious communities were tense and sometimes erupted into violence. The Maronite revolt in Lebanon in 1858/59 under the leadership of Ṭanyūs Shāhīn that began in the Kisrawān community, and the 1860 uprising of Christian peasants led by Yūsif Beg Karam against the large landowners, were two unique events in the Middle East that set the stage for the emergence of the Lebanese middle class. In response, the Ottomans inflicted a massacre

---

99 ففي ما رأيناه من نزاع اخيل واغاممنون وما سنراه من الوقائع المتوالية ولا سيما استطالت ترسيت على اغاممنون بعد ابيات من هذا
النشيد حجّة قوية على انّ الملِك لم يكن مستبدًا بأمره ورأيه بين اسحابه وإتباعه بل كان يشاورهم في الأمر كما فعل خلفاء العرب في صدر
ولقد زعم بعض الشرّاح استدلالاً بهذا البيت انّ هوميروس كانا يميل إلى الملِك الإستبدادي ... .الإسلام وكما نصّت الشريعة الإسلامية
وحقيقة الحال انّ اغاممنون لم يكن زعيم ملوك اليونان إلا اثناء الحرب لا قبلو ولا بعد ... المطلق وهو زعم تؤيّد فساده كل انشاد الإلياذة
وقد قام بأعباء قيادة الجند ورئاسة الدينية على ما يظهر من تولّيات شؤون العبادات كما كانت الخلافة والإمامة بيد واحد عند العرب
Bustānī, *Ilyādha*, 264 fn. 2; cf. Kreutz, *Arabischer Humanismus*, op. cit., 44. Bustānī on constitutionalism: وان اعلان .ولا شك أن مظاهر التواد والاخاء التي عمت البلاد ستكون أعظم ذكرى وأمتن أساس لهذا البناء الجديد
الدستور وتعميم المساواة يضمنان رسوخه Bustānī, *ʿibra wa-dhikrā*, op. cit., 158.

100 حان كلمة قالها شيخ الاسلام لجلالة السلطان يوم اعلان الدستور لجريدة بأن تنقش على صدر كل شيخ وقسيس، بل على صدر كل
مسلم ومسيحي بل على صدر كل عثماني وكل انسان. (...) Bustānī, *ʿibra wa-dhikrā*, op. cit., 160. This word was similar to Pascal's grain of sand with which he turned over what he called the political course (majran) of the world. It would be better for us (mā aḥrānā an) to take it as a slogan to be proud of […]به نتفاخر شعارا نتخذها أن احرانا فما .العالم سياسة مجري قال كما لها انقلب التي بسكال رمل بذرة اشبهها كان ما الكلمة تلك
(...) Bustānī, op. cit., 160 – 1.

101 وان هذا الدستور الذي شهد شيخ الاسلام بموافقته للشرع الشريف شهادة مزكاة بنصوص القرآن – والشورى حكم الاسلام – ليس
من ينكر موافقته لنصوص التوراة والانجيل Ibid., 161.

102 فلئن رأيناهم يوم اعلانه ملتفين حواليه يتعانق منهم الامام والقسيس والحاخام، يشهدون العالم أجمع على تآخيهم وتترقرق دموع
الفرح من مآقيهم (...) Ibid.

103 Weber, *Die jüdische Gemeinde im Damaskus des 19. Jahrhunderts*, op. cit., 47 – 8.

upon them that, in a twist of irony, finally paved the way for the autonomy (and, later, independence) of the Lebanon.[104]

Bustānī, who actively mitigated the sectarian tensions, feared a wildfire of violence in the Middle East. Important here is that the major cities in the Eastern Mediterranean had mixed populations, so that in Cairo alone we find Europeans, Greeks, Turks and Arabs, and therefore different religions in a relatively small area.[105] The Armenian question was looming. The riots of 1860 gave a foretaste of the genocide that would many years later come. However, Bustānī was not blind to Muslim suffering, which in his view had been greater than the suffering of the Armenians. In ethnic terms, he saw a striking difference between the Kurds and the Arabs on the one side, and the Turks on the other; for Bustānī, the Turks had also been the victims of other groups, and had more to complain about. Moreover, Bustānī stresses the fact that it was the Muslim leaders in Damascus who had protected the Christian population.[106] Bustānī praised the fact that, under Ottoman rule, Jews and Christians had enjoyed benefaction and liberty, which he deemed unparalleled in any Christian country.[107] All this stood at the core of the Ottoman "politics of harmony."[108]

The reason that Muslim suffering was greater than Christian suffering is at least partly due to the fact that, as Bustānī goes on to explain, the Muslims were in harmony with the conquering *umma*, which is the reason that they took all the burden of war, but also that, in contrast to the Christians, who were free from conscription, they had the advantage of forming the leadership. This caused anger among the Christians and little attempts were made to alleviate the situation. Even worse, many officers may have supported the incitement of latent hatred because they were keen to exploit this situation.[109] Bustānī gives lengthy explanations on sectarian conflicts that were due to ignorance and fanat-

104 Lewis, "The Idea of Political Freedom," op. cit., 272–3; Hanf, "Die christlichen Gemeinschaften," op. cit., 35; Axel Havemann 2002: 64, 72. Brockelmann, *Geschichte*, op. cit., 331. Kreutz, *Das Ende des levantinischen Zeitalters*, op. cit., 86–7.

105 Bustānī, *'ibra wa-dhikrā*, op. cit., 179–80.

106 Ibid., 155–6. اذا شكا الارمني لدم يُهْدَر ومال يُسْلَب فشكوى المسلم اعظم ibid., 155.

107 فان رؤساء الدين المسيحي والاسرائيلي، على فرض انهم لا ينظرون الا الى مصلحة انفسهم دون مصلحة أبناء دينهم ونعيذهم بالله من ذلك، فانهم بلا ريب يعلمون أن لهم في بلاد الدولة العثمانية من الميزة والنعم والحرية ما ليس لزملائهم شيء من مثله في جميع بلاد الدول المسيحية Bustānī, *'ibra wa-dhikrā*, op. cit., 162.

108 Ibid., 152.

109 ... إذا نظرت الى الدينين الغالبين وهما الاسلام والنصرانية، والى العناصر المختلفة التي يتألف منها هذا الجسم رأيت هناك أسبابا فالمسلم باتحاده بالدين مع الامة الفاتحة وقيامه دون المسيحي بعبء الحروب ورد الغزوات، لامتناع التجند .أخرى تدعو الى هذا الشقاق والجهلة وذوو الغايات من رجال الدين لا يدركون .والمسيحي يعد نفسه محكوما مظلوما .على المسيحيين، يرى له حق السلطة والسيادة والحكومة لاهية بمشاغلها بل ربما عمد كثريون من عمالها الى اثارة الاحقاد .كنه الغرض الواجب عليهم اداؤه بالتهوين على الفريقين الكامنة جرالمغنم يرجونه أو غاية يرمون اليها Ibid., 152.

icism. Both go along with each other; they live and die together. Besides religious, there is also racial fanaticism, which is no less strong.[110] Although the Lebanese Christians are deeply rooted in their country,[111] the power remained solely in the hands of the Muslims in general and the Turks in particular.[112] Therefore, ending discrimination is key to fighting religious fanaticism which means recruiting Christians alongside Muslims. Also, the best way to fight racial fanaticism is to spread the official language and to make the teaching of Turkish compulsory. These two strategies, together with the dissemination of knowledge and education, are designed to guarantee friendship and fraternity.[113] Consequently, for a constitutionalist like Sulaymān al-Bustānī, Sultan ʿAbdülḥamīd's abolition of the constitution,[114] when freedom was finally "expelled" from the country, must have appeared a major blow to decades of political efforts.[115]

We can hear voices of a distinct secularism in the Arab-Christian community. For example, that of Shiblī ash-Shumayyil (1850 – 1917), a Greek Catholic pioneer of the Nahḍa who fervently supported secularism. In his view, the separation of state and religion was a requirement for every form of social progress.[116] In a similar fashion, the publisher and journalist Jurjī Zaydān (1861 – 1914), born in Beirut into a Greek-Orthodox family, emphasized the universality of modern civilization, which stood above any religious justification.[117] With the foundation in 1454 of the Greek-Orthodox millet under Gennadios II, the Church lost the posi-

110 Ibid. والجهل رفيق ملازم للتعصب يعيشان ويموتان معا Ibid.
111 Ibid., 154.
112 ومَن مِنا ينكر أن الامة الاسلامية اعظم امم الدولة العثمانية بل (...) ان ددولة الجهل والتعصب قد انقضى ودالت دولة الفتن الدينية
فما احرى سائر ... فاذا كان الشيخ الاعظم المسلم التركي .ومَن ينكر ايضا أن الترك هم أرباب السلطة العظمى فيها .هي قوامها المكين
خدمة الدين من مسيحيين واسرائيليين وغيرهم أن يتسابقوا متهافتين الى احراز مثل ذلك المجد الباذخ (...) Ibid., 163.
113 ولكنه لو اتيح لنا أن نضيف رأيا الى تلك الآراء النيرة لقلنا أن أعظم الوسائل لضَمان اضمحلال التعصب الديني تجنيد المسيحيين
فان هاتين الوسيلتين، مع .وأعظم وسيلة لاضمحلال التعصب الجنسي تعميم اللغة الرسمية وجعل تعليم اللغة الركية اجباريا .مع المسلمين
تعميم اسباب العلم والتهذيب، يَضْمَنان توثيق عرى التواد والاخاء Ibid., 159.
114 ويقول .فلقد كانت لعهد مضى مطلقة يسرح المرء ويمرح ايان شاء، ويخالط شاء .ان اول ما يحرص عليه المرء حرية شخصه
ويعمل ما شاء، مما لا ينال سواه بأذى. ... كأن القسطنطينية رجعت الى زمن كاليغولا في رومة ... Ibid., 93.
115 Ibid., 131.
116 Shiblī ash-Shumayyil, *Majmūʿa* (Cairo: Muqtataf, 1900 – 10), vol. II, 296 – 7; ʿUmar Riyāḍ, *Islamic Reformism and Christianity: A Critical Reading of the Works of Muhammad Rashid Rida and His Associates, 1898 – 1935* (Leiden: Brill, 2008), 86; Kreutz, *Das Ende des levantinischen Zeitalters*, op. cit., 162.
117 Zaydān, *riḥla ilā Ūrubbā*, op. cit., 122 – 3, after Sharabi, *al-muthaqqafūn al-ʿarab wa-l-gharb*, op. cit., 56 (fns. 5 and 6). Aḥmad Fāris ash-Shidyāq founded *al-jawānib* in Istanbul in 1890, Buṭrus al-Bustānī *Sūrīya* in 1890, Salīm and Bishāra Taqlā *al-Ahrām* in Egypt in 1875, Adīb Isḥāq and Salīm Naqqāsh *al-Maḥrūsa* in Egypt in 1879, Fāris Nimr and Yaʿqūb Ṣarrūf *al-Muqtaṭaf* in Beirut in 1876, later moved to Egypt in 1886, Jurjī Zaydān *al-Hilāl* in 1892; v. Aghnāṭiyūs Dīk, *al-masīḥīya fī Sūrīa: tārīkh wa-ishʿāʿ* (Aleppo: Dar al-Kitab al-Muqaddas, 2008 – 9), vol. 3, 114.

tion that it had occupied within the Eastern Roman Empire, and yet it became an ally to the Ottoman state, so that church and state were tied again, even though the partnership was no longer equal.[118]

The close ties that the Church had to its own religious-ethnic community were in turn susceptible to a rising national consciousness that ignored the fact that the population of Southeast Europe was quite heterogeneous. According to Georg Ludwig von Maurer's 1835 account, Greece was after independence still in many respects the most heterogeneous country in Europe. In terms of society, "the kind and the degree of education is nowhere so wide-ranging as in the present-day kingdom of Greece";[119] and, in terms of religion, there were "Catholic settlers on islands such as Tinos, Syra, Naxos and Santorini that dated back to the age of the crusades and Venetian rule." Maurer noted that the relationship between the Catholics and the Greeks had once been tense.[120]

However, the widespread assumption that there was a rift along sectarian lines within Ottoman society is disputed. Current research suggests, however, that social consciousness tended to be attached to a certain milieu, so that at best some Muslim hatred was directed at the Druse, who were said to be disobedient in terms of the government.[121] This might be true for "social consciousness," but that does not rule out sectarian tensions. The Smyrna massacre of 1797, which peaked in the killing of some two thousands Greeks and the destruction of much of the foreign merchants' property, is a telling example. As was noted by the British consul in Smyrna, Francis Werry, this was triggered by a secular event, i.e., the provocative public display by the Venetians of democratic symbols connected to the French Republic's annexation of the Ionian islands in 1797.[122]

In contemporary Greece, the Poems of Ossian were used as a tool to create unity beyond sectarian lines, since they drew on a universal love for nature instead of on specific religious tenets. To the nineteenth-century reader, as O'Don-

118 Karpat, "Millets and Nationality," op. cit., 616. Dimaras, *Νεοελληνικός Διαφωτισμός*, op. cit., 226.

119 "Es existirt wohl kein Land in Europa, vielleicht keines in der ganzen bekannten Welt, in welchem so heterogene Elemente durch einander braussen, in welchem die Art und der Grad der Bildung so verschiedenartig ist, wie das heutige Königreich Griechenland." Von Maurer, *Das griechische Volk*, Zweiter Band [2nd vol.], op. cit., 22.

120 Ibid., 30–1.

121 Weber, *Die jüdische Gemeinde im Damaskus des 19. Jahrhunderts*, op. cit., 233.

122 Clogg, "The ‚Dhidhaskalia Patriki'," op. cit., 88.

nell (2015) argues, this proved an axiomatic truth,"[123] which must be understood in the context of the sectarian rifts and tensions at the time. The Poems of the supposed Gaelic minstrel Ossian, albeit an eighteenth-century fabrication either by the English writer James Macpherson or by Edinburgh *literati*, played an important role in European Romanticism, which was sustained by Rousseau and Herder, who sought the authentic literature of the ordinary people beyond the classical narratives.[124] In a similar vein, the Greek nationalist Thomas Paschidis wrote an article in French in 1886 on "The Turkish-Greek Question" in which he outlined a joint Turkish-Greek future based on the philosophy of Plato and Aristotle.[125]

This article seemed tolerant, but was in fact a manifestation of Greek supremacism. Ionos Dragoumis defined the main objective of contemporary Hellenism as lying in the attempt to become "the centre for the surrounding unredeemed tribes, i.e., the Arvanites, Armenians, Turks and Arabs,"[126] who would benefit greatly from the superiority of Greek culture. However, this was not going to happen, since, as Dragoumis complained, "the Great Idea," i.e., a Greek nation-state comprising all of Asia Minor, had already been given up, so that the political restoration of Hellenism was reduced to the unity of a nation within a state much smaller than the boundaries of the Byzantine Empire.[127] Greek national aspirations, according to Dragoumis, should be directed towards one Greece comprising all the islands and part of Macedonia and Epirus. Moreover, it would help the Albanians to create their own state, which would act as a bulwark against the enemies both of Albania and Greece.[128] The same Dragoumis

---

123 Kathleen Ann O'Donnell, "The Disintegration of the Democratic Eastern Federation and the Demise of its supporters 1885–1896 and the Poems of Ossian," in *ATINER'S Conference Paper Series* (Athens 2015): 3–14, here 14.

124 John H. Zamitto, "Die Rezeption der schottischen Aufklärung in Deutschland. Herders entscheidende Einsicht," in *Europäischer Kulturtransfer im 18. Jahrhundert. Literaturen in Europa – Europäische Literatur?*, ed. Barbara Schmidt-Haberkamp, Uwe Steiner and Brunhilde Wehinger (Berlin: Berliner Wissenschafts-Verlag, 2003): 113–38, here 120–2; Matl, op. cit., 350.

125 O'Donnell, "The Disintegration," op. cit., 11.

126 "Σκοποί τωρινοί τοῦ Ἑλληνισμοῦ: ... 2) Νά γίνη ὁ Ἑλληνισμός κέντρο γιά τίς γύρω ἀλύτρωτες φυλές (Ἀρβανιτῶν, Ἀρμενίων, Τοῦρκων, Ἀράβων)." Dragoumis, Ὁ Ἑλληνισμός μου καί οἱ Ἕλληνες, op. cit., chap. Κωνσταντινούπολη 1908, 131.

127 "Καταργεῖται τέλος πάτνων ἡ Μεγάλη Ἰδέα καί πολιτική ἀποκατάσταση τοῦ Ἑλληνισμοῦ εἶναι ἡ ἕνωση τῆς φυλῆς σ' ἕνα κράτος πιό συμμαζεμένο ἀπό τό βυζαντινό." Ibid., chap. Rome 1909, 141.

128 "Συζυτήσεις μέ τό Σουλιώτη γιά τό ζήτημα αὐτό καί ἀποκρυστάλλωμα τῆς δικῆς του καί τῆς δικῆς μου σκέψης. Τό τουρκικό κράτος ἔχει πρωτεύουσα τήν πρωτεύουσα τοῦ βυζαντινοῦ καί, ὅσο ὑπάρχει τό κράτος αὐτό, θά εἶναι καί ἑλληνική πρωτεύουσα ἡ Πόλη, προπάντων τώρα που ἔχει σύνταγμα (ἤ δῆθεν σύνταγμα) ἡ Τουρκία. Μά οἱ ἀγῶνες μας ὅλοι πρέπει νά γίνουν

who had called Demotic Greek an "illness" (ἀρρώστια, v. 122) now glorified it as a "symbol of the true rebirth of Hellenism."[129] Dragoumis' role model was Italy, a "nation divided into many smaller communities" with each community forming its own state while being in a constant state of conflict with surrounding states. This produced many individuals with "invigorated characters, with political gumption, with honed minds," and therefore slowly attracted people of "all possible qualities of character and mind."[130] This sounds like an idealized version of reality, but Dragoumis is not content with a situation where people give up on greater ideals, a situation where they will later rest, have a good time, forget, sleep, not toil, and dream of Plato's Republic, where they will live in a socialist manner while making as little effort as possible. In Dragoumis' eyes, such is the situation of contemporary European individuals.[131]

Dragoumis believed that Greece could become more than just a small country in Southeast Europe. The consulates in Turkey would be great tools for Greek policy if the country were aware of its political capabilities. At the same time, he continues, this power poses a risk because the panhellenic programme is becoming increasingly exposed to the Turks and other foreigners who associate it with the Greek state – a state that is otherwise helpless and unable to make imperialist politics because it lacks an army, a navy and statesmen worthy of the name.[132]

ἀγῶνες γιά μιάν Ἑλλάδα, που θά ἔχει ὅλα τά νησιά καί ἕνα μέρος τῆς Μακεδονίας καί τῆς Ἠπείρου καί που θά βοηθήση καί τούς Ἀρβανίτες νά κάμουν τό κράτος τους ἀπό πάνω ἀπό τήν Ἑλλάδα, προπύργιο ἐνάντια στούς τρίτους." Ibid.

129 "[...] Δημοτική γλῶσσα σύμβολο ἀληθινῆς ἀναγέννησης γιά τόν Ἑλληνισμό." Ibid., 142.

130 "Ἕνα ἔθνος μοιρασμένο σέ πολλές μικρότερες κοινωνίες, που ἡ καθεμιά τους εἶχε ξεχωριστό πολιτικόν ὀργανισμό καί ἦταν κράτος καί κάθε κράτος μάλλωνε ἀδιάκοπα μέ τά τριγυρινά του, βγάζει πολλά ἄτομα μέ τονωμένους χαρακτῆρες, μέ πολιτική ὀξύνεια, μέ μυαλό ἀκονισμένο καί φτειάνει σιγά σιγά τούς ἀνθρώπους, που μαζεύουν μέσα τους [...] ὅλες τίς δυνατές ἰδιότητες τοῦ χαρακτήρα καί τοῦ μυαλοῦ [...]." Ibid., 148.

131 "[...] Καί ὕστερα ξετεντώνονται [stretched again], χαλαρώνονται, ξεκουράζονται, καλοπερνοῦν, ξεχάνονται, κοιμοῦνται, βλέπουν, συλλογίζονται, προπάντων δέν κοπιάζουν καί ὀνειρεύονται πολιτεῖες τοῦ Πλάτωνα, ὅπου θά ζοῦν σοσιαλιστικά μέ ὅσο μποροῦν λιγώτερο κόπο καί ὅσο εἶναι δυνατόν περισσότερη καλοπέραση. Σέ μιά τέτοια κατάσταση εἶναι τώρα οἱ Εὐρωπαῖοι ὡς ἄτομα. [...]" Ibid., 148–9.

132 "Τό ἑλληνικό κράτος ἀπόχτησε μέ τά προξενεῖα του στήν Τουρκία δύναμη, που δέν τή συναισθάνεται. Ἡ δυναμή αὐτή εἶναι κίνδυνος, γιατί ἀδιάκοπα ἐκθέτει στούς Τούρκους καί σ' ἄλλους ξένους τό πανελληνικό πρόγραμμα τοῦ Ἑλληνισμοῦ που τό θεωροῦν αὐτοί ὡς πρόγραμμα τάχα τοῦ κράτους, κράτους ἀνίσχυρου κατά τ' ἄλλα καί ἀνίκανου νά κάμη ἰμπεριαλιστική πολιτική, γιατί στερεῖται στρατό καί στόλο καί πολιτικούς ἄνδρες ἄξιους τοῦ ὀνόματος." Dragoumis, Ὁ Ἑλληνισμός μου καί οἱ Ἕλληνες, op. cit., Κωνσταντινούπολη 1908, 136.

Also, as Dragoumis argues, the Halki Seminary (theological school) has to educate people with the purpose of saving the Greeks, and not with the purpose of growing rich, living well, and becoming fat. He accuses the Metropolites of indulging in a pleasant way of life and of avoiding work.[133] Dragoumis then unleashes an invective against the Church, whose exploitation of the Greek population dates back to the Eastern Roman ("Byzantine") era. He accuses the bishops of sacrificing everything in order to enjoy the protection of the more powerful, which makes them something of an alien element: "They are no Greeks; they are Christians and Epicureans."[134]

We return to religion at this point, since Dragoumis sees Christianity as "inseparable from our history" and its national continuation. Although he has little interest in religion as such, since "all religions are the same" and are therefore interchangeable, he confesses to going to church, to making the sign of the cross, to lighting a candle, and to putting a coin into the collection bag before standing for an hour or two in an attitude of devotion. Perhaps, he muses, Christianity does have some assets, "which is why we are used to it."[135] Dragoumis' attitude in terms of Christianity is somewhat lukewarm. In his view, the religion is more a useful tool than something to be appreciated for its spiritual dimension. He concludes that the church is more a political factor with which one has to find the

---

133 "Ἡ θεολογική σχολή τῆς Χάλκης νά βγάζη ἀνθρώπους μέ σκοπό νά σώσουν τόν Ἑλληνισμό, ὄχι μέ τό σκοπό νά πλουτίσουν, νά καλοζήσουν, νά παχύνουν, νά καλοπεράσουν. Πῶς μποροῦν νά ἐργασθοῦν καλά οἱ μητροπολῆτες, ἀφοῦ σκοπός τους δέν εἶναι νά ἐργασθοῦν καλά, ἀλλά νά ζήσουν εὐχάριστα. Τέτοια εἶναι ἡ ἀνατροφή τους." Ibid., Philippopolis [Plovdiv] 1904, chap. B, § 1: 37–8.

134 "Τά ἀπομεινάρια τοῦ βυζαντινισμού σχετίζονται μέ μένα τόν νέον Ἕλληνα καί ὅπως σέ κάθε περίσταση οἱ Βυζαντινοί κοίταζαν νά ὠφελιοῦνται ἀπό τούς δυνατωτέρους τους μέ κολακεῖες καί πονηρίες, ἔτσι καί τῶρα ἐμᾶς τούς Ἕλληνες τῆς Ἑλλάδος μᾶς ἐκμεταλλεύονται, μᾶς κολακεύουν, ἐπειδή ἔτυχε νά ἔχωμε κάποια δύναμη στά χέρια μας καί κάμποσο χρῆμα. Ἄν δέν εἴμαστε μεῖς. [...] Μποροῦν νά θυσιάσουν ὅλα οἱ δεσποτάδες γιά νά ἔχουν τήν προστασία τῶν δυνατωτέρων τους. Οἱ δεσποτάδες δέν εἶναι Ἕλληνες, εἶναι χριστιανοί καί ἐπικούρειοι." Ibid., Philippopolis [Plovdiv] 1904, chap. B, § 4: 39.

135 "Ὑποστηρίζω τήν θρησκεία μας, ἐπειδή εἶναι ἀχώριστη ἀπό τήν ἱστορία μας, εἶναι ἡ συνέχεια τῆς ἱστορίας τοῦ γένους. Καί ὑποστηρίζω τήν θρησκεία μας, καί ἐπειδή, ἄν δέν τήν εἶχαν αὐτήν, θά εἶχαν ἄλλη καί ὅλες οἱ θρησκεῖες εἶναι τό ἴδιο πράγμα. Ἄν ἔπαιρνα ἀπό μερικούς Ἕλληνες τή θρησκεία τους, θά γίνονταν ἤ ἄθεοι ἤ μασόνοι ἤ δῆθεν φιλόσοφοι ἤ θά ἔπαιρναν ἄλλες ἀνόητες πόζες καί ἰδέες τωρινές πού τίς σιχαίνομαι. [...] Καί τήν θρησκεία μας τήν ὑποστηρίζω πηγαίνοντας στήν ἐκκλησία, κάνοντας τόν σταυρό μου, ἀνάβοντας ἕνα κερί, δίνοντας τήν πεντάρα μου στόν δίσκο, στεκόμενος μιά δυό ὧρες ὄρθιος. Ἴσως ἔχει καί τό καλό ἡ θρησκεία μας, πως τήν ἔχομε τόσο συνηθισμένη, που δέν εἶναι ἀνάγκη νά τή συλλογιζόμαστε." Ibid., Κραβασαράς Λοκρίδος, 21 May 1904, 78–9.

modus vivendi.[136] Yet, he feels committed to Greek Orthodoxy, which is the reason that he admits to kissing icons and the hands of priests, and to generally supporting the church system, for it was the church that had helped subjugated Hellenism to survive. It was due to the church that the Greek language, as the essence of Greek culture, had been kept alive. Nowhere but in the church did Greeks gather and it was the only place where they celebrated their status as one nation.[137]

In contrast, Faraḥ Anṭūn (1874–1922), another Greek-Orthodox thinker on the other shore of the Mediterranean, advocated a strong separation of religion and the state. Born in the Lebanese town of Tripoli, he spent most of his life in Egypt, which was out of reach of the Ottoman sultanate and its oppressive regime. In Egypt, he was heavily influenced in his call for a society in which all religions were equal and overarched by a secular nationalism by Ernest Renan and Gustave Le Bon, whose popular volume on the *Psychology of the Masses* (*Psychologie des foules*, 1895) was one of the first Western European books to be translated into Arabic.[138] His 1903 treatise on the medieval Muslim philosopher Averroes (Ibn Rushd, d. 1198), *Ibn Rushd wa-falsafatuhu* (Averroes and his philosophy), caused an outrage since he presented Averroes as pioneering the separation of religion and state. One of Anṭūn's fiercest critics was the Azhar Sheikh Muḥammad ʿAbduh.[139]

Secularism, as defined by Faraḥ Anṭūn, means "separating civil rule from religious rule." However, Arab culture and society are strongly associated with Islam, which makes separating them a difficult task.[140] What makes a secular state in the Arab world? Muḥammad ʿAbduh wondered whether it was sufficient for the *umma* to choose a ruler within the framework of a secular, i.e., democratic, process, so that the ruling power became a civil power, or whether the politics

---

136 "[...] ἡ ἐκκλησία, εἶναι ἀνεκτίμητη δύναμη, ὀργανισμένη ἀπό δῶ καί πολλούς αἰῶνες. Εὐρεθήτω τό *modus vivendi.*" Ibid., Δεδέαγατς [= Alexandroupoli – Dedeağaç] 1905, 86.

137 "Ὑποστηρίζω τήν ὀρθόδοξη θρησκεία, ἐπειδή εἶναι ἑλληνική. Καί τήν ὑποστηρίζω πῶς; Πηγαίνοντας στήν ἐκκλησία, κάνοντας τό σταυρό μου, βοηθῶντας τοῦς δεσποτάδες στά χρέη τοῦς, φιλῶντας τίς εἰκόνες καί τῶν παπάδων τά χέρια, δίνοντας τή δεκάρα μου στούς δίσκους. Ὑποστηρίζω τό σύστημα τῆς ἐκκλησίας. Στόν ὑπόδουλο Ἑλληνισμό ἡ ἐκκλησία εἶναι ἕνα κέντρο καί δέν ξεχωρίζει ἀπό τό σχολεῖο. Πνευματική ζωή σέ πολλά μέρη καί γιά πολύν καιρό, ἑλληνική, δέν ὑπῆρχε ἄλλη ἀπό τίς λειτουργίες στήν ἐκκλησία καί ἀπό τό μάζωμα τοῦ κόσμου ἐκεῖ. Ἦταν τό μόνο ἑλληνικό ἀνθρωπομάζωμα, ἡ μόνη ἐθνική διαδήλωση. Καί ἀκόμη εἶναι, σέ πολλά μέρη. [...]." Ibid., 88–9.

138 Kreutz, *Arabischer Humanismus*, op. cit., 79–80.

139 Dominique Urvoy, *Ibn Rushd (Averroes)* (London and New York: Routledge, 1991), 120–1. Hourani, *Arabic Thought*, op. cit., 148; Kreutz, *Das Ende des levantinischen Zeitalters*, op. cit., 163.

140 Barqāwī, *muḥāwala*, op. cit., 84.

of the ruler had to be of a civil nature. If the task of the ruler was to implement the Islamic *sharia*, then he is a religious ruler even though he is the product of a civil (secular) election. In his discussion of Faraḥ Anṭūn's line of argumentation, Muḥammad ʿAbduh made clear that whoever ruled the *umma* could not refrain from religion.[141]

Salāma Mūsā (1887–1958), a public intellectual and between 1924 and 1944 publisher of the journal *al-majalla al-jadīda* (*New Journal*), followed a more radical approach concerning the relationship between religion and the state. His point of departure was the Islamic conquest of Egypt in 639, which he regarded as a historical catastrophe for the progress of the region. The idea conceived by the Prophet Muḥammad of a deliberative assembly (*majlis shūrawī*), which had so often been praised by contemporary intellectuals as Islam's original contribution to the reign of law, was, according to Mūsā, nothing but an illusion. Mūsā advocated instead a future Egypt that would be part of Europe, and rejected any romantic turn to the Islamic past.[142]

Secularism was in a way a pivotal idea in Mūsā's thought, and he tried to assess it against the background of Islamic history in his 1910 book *Introduction to the Super-human* (*muqaddimat al-subirmān*).[143] Here, Mūsā defines secularism (*ʿilmānīya*) as something that unfetters the educational system and political power from the shackles of religion in the light of science (the *ʿilm* in *ʿilmānīya*), which he regarded as being an antithesis to religion. Mūsā argues that, due to modern science, a secular education unhampered by the tenets of religion has arisen and now enables mankind to discuss freely issues such as marriage and divorce, family, property and the beginnings of the world in a manner that does not take even God into consideration.[144]

Looking at the "first humanist movement," Mūsā praises its role in preparing the ground for the independence of the human mind and its crafting of a non-religious literature whose authors did not hesitate to draw on pagan Greek and Latin sources. This was the starting-point of the reformatory idea that

---

141 Ibid., 59.

142 Gershoni / Jankowski, *Egypt, Islam, and the Arabs*, op. cit., 112–3, 115; cf. Kreutz, *Das Ende des levantinischen Zeitalters*, op. cit., 277–8.

143 ʿAbdarraḥmān Shahbandar, "at-taṭawwur al-ijtimāʿī wa-s-siyāsī l-ḥadīth fī sh-sharq al-adnā [1931]," in idem, *al-maqālāt*, ed. Muḥammad Kāmil al-Khaṭīb (Damascus: Manshurat Wizarat ath-Thaqafa, 1993 [first publ. in *al-Muqtaṭaf* 2:79]: 335–48, here 337; cf. Kreutz, *Das Ende des levantinischen Zeitalters*, op. cit., 282.

144 Salāma Mūsā, *muqaddimat as-subirmān* [1910] (Cairo and Alexandria: Dar al-Mustaqbal, 1977), 5–6. Bassam Tibi, *Vom Gottesreich zum Nationalstaat: Islam und panarabischer Nationalismus* (Frankfurt/Main: Suhrkamp, 1987), 82–4. Kreutz, *Das Ende des levantinischen Zeitalters*, op. cit., 277–8.

every believer was free to communicate with God without intermediary institutions. This morphed into a religious movement led by Martin Luther, who understood that freedom of conscience would finally undermine the authorities of state and religion, which made it a useful tool to fight repression.[145]

Mūsā goes on to explain that this was not what Luther had intended exactly. Rejecting the revolution of the peasants, Luther defended the rights of the lords and the nobility and, in doing so, according to Mūsā, he also initiated the idea of free economic endeavour, as well as of free competition between individuals. This is not entirely accurate, since the origins of free markets are to be found instead in the northern Italian cities of the fourteenth century. But here we wish to follow Mūsā's argumentation, while bearing in mind that he is a Marxist. As a Marxist, he praised historical development up to the point when modern governments, he says, began to adopt socialist positions. He stresses that freedom of conscience stands at the beginning of modern scientific empiricism.[146]

Both Erasmus and Luther, "the German reformer and the leader of Protestantism," were actually propagandists for what Mūsā calls "religious democracy."[147] Moreover, Luther's rise was accompanied by the emergence of different European nationalisms, which were fostered by his translation of the Bible into German, which would then become a literary language.[148] This is correctly observed, and Mūsā's conclusion is worth thinking about: "Had the Europeans placed religion and the language of religion over nationalism, Europe would by now be one country and Rome its capital."[149] In contrast, the idea of the nation-state is still little rooted in the Islamic world, which helps explain the fact that in most Islamic countries religious identity outweighs national identity.[150]

As a Marxist, Mūsā follows the "economic interpretation of history," and believes that many phenomena in philosophy, religion and literature can be better explained through reference to economic conditions.[151] He therefore seeks human progress and praises Bacon and the Greek philosophers for thinking in the abstract,[152] as well as Descartes for casting philosophical doubt on every-

145 Mūsā, *mā hiya an-nahḍa?*, op. cit., 58.
146 Ibid., 58; Kreutz, *Zwischen Religion und Politik*, op. cit., 175.
147 Mūsā, *mā hiya an-nahḍa?*, op. cit., 67.
148 Ibid., 111.
149 Ibid., 113.
150 Cf. Cook, *Ancient Religions*, op. cit., 27; Kreutz, *Zwischen Religion und Politik*, op. cit., 33–4, 166. Karsh/Karsh, *Empires of the Sand*, op. cit., 171–2.
151 Mūsā, *mā hiya an-nahḍa?*, op. cit., 59.
152 Ibid., 78.

thing around us.[153] All of this is manifested in European culture, which is based on experience and reason,[154] and which peaks in its independence from the church, which is something that Machiavelli and others had advocated.[155] But, how do these concepts fit into a Muslim context? Can they provide a blueprint for Muslim reform?

There had in fact been a process of separation between governmental and religious institutions since Muḥammad's death, a process that had taken more than 400 years.[156] A leading Syrian nationalist, ʿAbdarraḥmān Shahbandar, argued in a 1931 article that the Ḥamidian era, when religious and worldly power merged, with the sultan becoming "God's shadow on earth," was similar to some periods in Catholic Europe.[157] However, when the tenth-century Islamic domain fragmented into smaller sultanates – which had hitherto been *de facto* secular institutions – they officially gained their legitimization from serving Islam, while the idea of the caliphate – although strongly opposed by zealot movements – underwent a parallel process of secularization.[158] Finally, the caliphate became under ʿAbbāsid rule a "model of world order" while representing a Muslim elite whose declared goal was the toppling of non-Muslim societal systems.[159]

The famous Egyptian scholar Ṭāhā Ḥusayn, himself a secularist thinker, once argued that the "despotic" east in a sense gained victory over the "enlightened" west when the latter was victorious in military terms, but vanquished on the spiritual front – this is an oft-heard criticism that Muslims make of the West, and indicates how difficult it is for secularism to take root in Islamic societies.[160] When tensions rose in Egypt due to the predominance of the Turko-Albanian aristocracy, and when a new nationalist and egalitarian order loomed, Ḥusayn

153 Ibid., 81.
154 Ibid., 83.
155 Ibid., 111.
156 Lapidus, "Islamisches Sektierertum," op. cit., 167.
157 Kreutz, *Das Ende des levantinischen Zeitalters*, op. cit., 283; cf. Karsh/Karsh, *Empires of the Sand*, op. cit., 171; cf. Weingrod (1990): [...] once religious nationalist ideologies emerge, they dovetail with belief in the efficacy of saintly powers." Alex Weingrod, "Saints and Shrines, Politics and Culture: A Morocco–Israel Comparison," in *Muslim Travellers: Pilgrimage, Migration, and the Religious Imagination*, ed. Dale F. Eickelman and James Piscatori (London: Routledge 1990: 217–35, here 233.
158 Lapidus, "Islamisches Sektierertum," op. cit., 170–1.
159 Ibid., 167, 169, 179.
160 كان الغرب منتصراً من وجهة العسكرية ولكن الشرق كان ينتصر من وجهة العقلية الشعورية Ibid., 277–8. On Ṭāhā Ḥusayn and secularism cf. Bassam Tibi, *Islamischer Fundamentalismus, moderne Wissenschaft und Technologie* (Frankfurt/Main: Suhrkamp, 1992), 56.

took Plato as a negative example of an elitist philosopher who, although an opponent of the aristocracy and a supporter of egalitarianism, did not trust society to organize itself. This is what Ḥusayn saw as being manifest in contemporary Egypt: an elite detached from society and a society in which the individual was largely absent.[161]

In Ḥusayn's account, Platonist philosophy, if put into practice, would abolish the institution of the family. Following generations would then no longer be the children of their parents, but instead the children of the government, which would take care of them and raise them until they reach the age when they have to do military service. Whether this account is accurate or not, Ḥusayn praised the Athenian model of democracy and its liberation of the individual as an antidote to authoritarian concepts of society.[162] This is why Ḥusayn champions Aristotle's ideas which provide a sustainable foundation for modern-era rationalism and humanity's journey to progress.[163] Although societies have undergone certain changes and human dignity ranks higher than in the past, Aristotelian thoughts have much to offer present-day societies now that slavery has been abolished, but social exploitation endures. Ḥusayn's plea for an Aristotelian society that safeguards individual freedom has never been realized in the Islamic Middle East.[164]

3.3 Eastern anti-Westernism

According to Ghalioun (2007), a civilization is a means of producing and developing culture, which is why a term such as "Western civilization" is justified as it is primarily a secular entity, while a term such as "Arabic-Islamic civilization" reflects the superior role of religion. In contrast to the Arabic-Islamic world, the West was shaped by its neutralization of religion, so that monotheism appeared to be increasingly alien to the Western spirit. Instead, the West tried to find its own foundation in the pre-monotheistic era, i.e., the era of ancient Greece and its heritage.[165] This sounds familiar, and is indeed very much in line with the ideas of Salāma Mūsā.

Nevertheless, the West, according to Ghalioun, cannot deny that it is massively influenced by Christianity, which makes it a Christian civilization after

161 Ḥusayn, *qādat al-fikr*, op. cit., 246–247.
162 Ibid., 248; Kreutz, *Arabischer Humanismus*, op. cit., 61.
163 Ibid., 259.
164 Ḥusayn, *niẓām al-Athīniyīn*, op. cit., 317–9.
165 Ghalioun, *naqd as-siyāsa*, op. cit., 245.

all. All its basic values and knowledge can be traced back to Christianity. Since the beginnings of its own Christianization in the ninth century, the West has liberated itself from barbarism and savagery (*hamajīya*) by means of Christianity, which is the reason that the West finally sees itself as a rational civilization (*madanīya ʿaqlānīya*) with secularism as one of its most powerful values. However, what is key to this process are not certain elements within Christianity, but rather a general commitment to its values.[166]

In this respect, Islam, Ghalioun goes on to argue, does not much differ from Western civilization. As a form of cultural belonging, Islam is not necessarily tied to belief, and nor is it the driving force behind sectarian conflicts. As a melting pot, Islamic civilization used to encompass different cultures, religions and peoples, and blended what was already there in terms of intellectual, rational and ethical elements. In this sense, there is not one unique Western civilization, but different cultures, each of which belongs to that civilization. Civilization, in turn, is a higher form of humanity that human communities seek to create and that they use to transform themselves into civilized societies.[167]

Huntington (1996) has argued that "[t]he central elements of any culture or civilization are language and religion," which is the reason that civilizations are surprisingly durable entities.[168] Ghalioun points out in a similar vein that Arabic-Islamic civilization came into being through Islam and, although it embraced values of universal meaning, it blended them together with Arabic culture, the source of its own genesis.[169] However, the civilization of the Arabs has been in decline for several centuries, so that now it is nothing more than a "stiff corpse."[170] Ghalioun rejects the popular explanation that this is the outcome of abandoning or neglecting heritage (*turāth*), because, in his mind, this situation was inevitable given that senescence is the fate of all civilizations. From this point of view, reviving a civilization already in decline was a hopeless endeavor.[171]

The fate of great cultures is not, Ghalioun goes on to explain, to die out, but rather to withdraw into their shells and preserve the remains of their power in the hope of a future opportunity. When Western civilization became a new source of global norms and values, a civilizational formula for organizing society, Arabic culture refused to take up the challenge and refrained from giving answers to

166 Ibid.
167 Ibid.
168 Huntington, *The Clash of Civilizations*, op. cit., 59.
169 Ghalioun, *naqd as-siyāsa*, op. cit., 245.
170 Ibid., 237–8.
171 Ibid., 248.

a wide range of issues affected by the Western challenge. According to Ghalioun's analysis, Arabic culture behaved in the same way as Western civilization had during the medieval era, when it encountered an Islamic civilization that was much more progressive than it is today.[172]

The Western response to the Islamic supremacy of the past was to wage a propaganda war against Islamic civilization, a war that, according to Ghalioun, has left its mark on the Western psyche today. It is the source of the hostile attitude towards the Arabs, whom the West regarded as its main competitor in terms of historical claims about contributions to the culture of the Mediterranean, which is the reason that the West tries to belittle the achievements of Islam and the Arabs. Over the course of several centuries, then, this hostility has turned into a permanent attack.[173]

The most important phenomenon in this context had been the crusades, which marked a major stage in the West's rise to intellectual and spiritual dominance, while also seeing the attempt to eradicate the ancient Oriental influence on the area. In Ghalioun's account, the failure of the crusades did not change Europe in its attempts to rid itself of any Islamic and Eastern influence. Thus, the crusaders' spirit kept glowing until the *Reconquista* took place. This was the war that encouraged Europe to open a new chapter in history and to gain mastery in the intellectual realm, too.[174]

Ghalioun's view might be very popular, but it misrepresents the facts. In Europe, the crusades had always been an issue of dispute, and were sometimes viewed as an example of medieval zealotry and bigotry. For example, the Scottish writer Sir Walter Scott portrayed the crusaders in the nineteenth century as "backward and unenlightened, crudely assailing more civilized and sophisticated Muslims." Also, the ostensible grievance among present-day Muslims is a modern invention, and enjoys little credibility vis-à-vis the fact that it was the Seljuk and Fāṭimid victories over the Arab lands that had opened the doors for the crusaders.[175] Moreover, the impact of the crusades on the Muslim public was limited up until the nineteenth century, when a French history of the crusades was translated into Arabic. Before that, the Islamic world had viewed the crusades within the framework of the old border wars with the Eastern Roman Empire. The Mongolian invasion in the thirteenth century, and the

172 Ibid., 250.
173 Ibid., 251.
174 Ibid., 251–2.
175 Jonathan Riley-Smith, *The Crusades, Christianity, and Islam* (New York: Columbia University Press, 2008), 65–9.

Black Death and other catastrophes in the fourteenth century, were of much greater significance.[176]

The French history book that was translated into Arabic in the nineteenth century was Joseph-François Michaud's *Histoire des Croisades* (1812–22). Since the Arabic language lacked a word for the crusades, the term *ḥarb aṣ-ṣalīb* (*al-ḥurūb aṣ-ṣalibīya* is more common today) was coined and quickly became part of Middle Eastern political discourse. The book's translator was Maximos Maẓlūm III, the Melkite (Greek-Catholic) patriarch of Antioch, Jerusalem and Alexandria, who had his own anti-Protestant agenda. The translation instilled fears of Western invasion into the Arab public because Michaud, in contrast to Walter Scott, took a clear stance in favor of the crusades and against Muslims.[177] This has had a massive influence on the Muslim view of the crusades while, as Riley-Smith (2008) points out, the West "has not tried to counter the Muslim reading of history" – also in terms of *jihād*.[178] The crusades became harnessed in the second phase to the pan-Arab cause, although it was the Turks, and not the Arabs, who had expelled the Frankish invaders. Since then, and despite the historical facts, Arab unity has been identified as the root cause of the victory over the crusaders. From this point of view, any attempt, by whichever side, to organize the Arab world according to the concept of nation-state, is seen as a way of weakening its unity.[179] Theories of when and how this unity came to an end, if it ever existed, usually focus on the last days of the Ottoman Empire. In fact, the role of the Western powers in the decline of the Ottoman Empire is ambivalent.

As Barqāwī (1988) has shown, the Arabs stood with the West when the latter was at war with the Ottoman Empire in 1914, which raises the question of whether Arab nationalism is nothing but part of the West's attack on the Ottoman Empire.[180] However, this would be a somewhat simplistic view against the backdrop that in the nineteenth century the European powers rather supported the Ottoman Empire against Russian expansion, and only later sided with Italy when it was about to annex the Ottoman province of Libya.[181] Although the fall of the Ottoman Empire was self-inflicted in a world that was turning into an order of nation-states, Arab nationalism never made it beyond an elite project

---

176 Endreß, "Der Islam und die Einheit ...," op. cit., 281.
177 Iris Shagrir and Nitzan Amitai-Preiss, "Michaud, Montrond, Mazloum and the First Historyof the Crusades in Arabic," in *Al-Masāq*, vol. 24, no. 3, December 2012: 309–12, here 311–2.
178 Riley-Smith, *The Crusades*, op. cit., 76.
179 Efraim Karsh, *Islamic Imperialism: A History*, op. cit., 133–4.
180 Barqāwī, *muḥāwala*, op. cit., 79.
181 Clark, *Sleepwalkers*, op. cit., 250.

to which the Arab masses were almost complete strangers. Strongly influenced by Western European culture, the nationalist movement in Damascus was restricted to intellectuals and students.[182]

The Arab elite was therefore reluctant to give up on the Ottoman Empire and only half-heartedly tried to gain their own nation-states. As Barqāwī has pointed out, Adīb Isḥāq (1856–1885), although a great admirer of the French philosophers of the Enlightenment, remained an "Ottoman" against Ottoman rule. Even Faraḥ Anṭūn, one of the greatest supporters of a separation of religion and the state, said that "we are the true Christians and our religion never intervened in politics. We are not responsible for the actions of Western Christianity and our loyalty is with the East. We were always loyal to the sultan."[183] The main aim of the Arabic nationalist movement was to move away from Turkey and closer to Europe, i.e., Western Europe, the hotspot of modernity. It is for this reason that Arabic nationalism is accompanied by a defense of Western Europe, and especially of France and England. It is no accident that the Europeanization and affirmation of the West emerged from rich Beirut merchants by the names of Michael Mudawwar, Francis Marrāsh, Louis aṣ-Ṣābūnjī, and Rizqallāh Ḥassūn.[184]

During the 1913 Arab Congress in Paris, which aspired to re-define the status of the Arab countries vis-à-vis the center, one of its speakers, ʿAbdalghanī al-ʿUraysī, declared that the Arab peoples did not think about a separation as long as the Ottoman state had a working constitution worthy of its name and guaranteed the rights of all its subjects.[185] Iskandar Bey ʿAmūn, deputy of the Non-Centralist Party, said that the Arab *umma* does not wish to split off from the Ottoman State, but wishes instead to have a rule that is more appropriate to every single nation under its rule, which is the reason that the party favors a real Ottoman government over a Turkish or an Arab one. Similarly, ʿAzīz al-Miṣrī, opponent of the Paris Conference, expressed his hope that there would be an Arabic-Turkish Empire styled after the model of the Austro-Hungarian monarchy.[186]

Although the Paris Congress contributed immensely to strengthening the Arab nationalist idea, separation from the Ottoman Empire or an independent Arab State was never proposed.[187] The Arab nationalist discourse was concerned primarily with proving equality between Arabs and Turks, and arguing that the

---

182 Karsh, *Islamic Imperialism*, op. cit., 127–8. Arsuzi-Elamir, *Arabischer Nationalismus*, op. cit., 235.
183 Barqāwī, *muḥāwala*, op. cit., 72.
184 Ibid., 70.
185 Ibid., 73.
186 Ibid., 74–5.
187 Ibid., 73.

Arab *umma* would not accept being compelled to subscribe to a different nationalism. As a result, the general attitude at the Paris Congress was that of an oppressed *umma*, emphasizing the unity of the Ottoman State and the Arab right to political equality.[188]

This implies that the concept of the Arab *umma* remained vague in the minds of Arab nationalists at that time. The term could encompass different peoples, sometimes bound together by religion, sometimes by a common state, sometimes by language. Therefore, it was blended with other terms such as community, people, race, state and religion.[189] Arab awareness of its ethnic identity took place at a time when the Ottoman State was crumbling and it was trying to survive by turning to reform – to no avail. Even its religious foundation was questioned.[190] Pan-Arabists such as Ṣāṭiʿ al-Ḥuṣrī denied that it was Islam that constituted Arab unity; rather, he viewed Arab unity as a requirement for Islamic universalism.[191]

If we assume that Christians were naturally more inclined towards secular Arab nationalism than Muslims, Barqāwī goes on to argue, then we must ask where Turkish nationalism comes from if not from Muslims. Since Christians were largely absent from the Turkish national movement, secular nationalism cannot be merely a Christian project.[192] This must also be true for the Arab world: in this sense, the idea of nationalism was initiated by Christian thinkers, but advanced by Muslims.[193] We therefore find three formulas for Arab nationalist thought before the First World War, i.e., an Arab-Ottoman nationalism, an Arab-Islamic nationalism, and a secular Arab nationalism.[194]

Finally, Barqāwī might be right when he states that the Arab nationalist consciousness emerged and developed through confrontation first with Ottoman rule, second with Turkish chauvinist nationalism, and only third under the influence of Western ideas.[195] Arabic nationalism was forged between the two world wars, and was shaped further after the Second World War.[196] The ruling elite in Damascus found a catalyst for their cause when the district of Alexandretta

188 Ibid., 75–6.
189 Ibid., 77.
190 Ibid., 69.
191 Karsh, *Islamic Imperialism*, op. cit., 133; cf. Karsh/Karsh, *Empires of the Sand*, op. cit., 171–2.
192 Barqāwī, *muḥāwala*, op. cit., 71.
193 Ibid., 78.
194 Ibid., 89–90.
195 Ibid., 78.
196 Ibid., 67.

(Hatay) fell into the hands of the Turks. As a consequence, they turned to the Palestine question and placed "Zionist danger" on the pan-Arab agenda. In doing so, they tried to corroborate their legitimacy as a proponent of Arab interests, while at the same time creating a distraction from the Alexandretta question.[197]

The assumption is questionable against this background that secular Arab nationalists once supported the revolution against the Turks and sided with the Sharif, Ḥusayn bin Ali, but turned into an anti-Western movement due to the colonization of Iraq and Syria, and to Balfour's promise to the Zionist movement.[198] As Karsh (2006) has shown, Ḥusayn and the Hashemites never delivered on their part of the bargain, and created the myth of Western betrayal despite the fact that they had gained vast swaths of land.[199] The sense of being a victim contributed to a longstanding inferiority complex vis-à-vis Western Europe.[200]

Nazi propaganda added to anti-Western resentment. In Egypt, such propaganda presented the Second World War as a war of liberation against British imperialism.[201] In 1942, it declared Nazism and Islam to be twin ideologies that were naturally opposed to British imperialism.[202] Four years later, Ḥasan al-Bannā, founder of the Muslim Brothers, praised Ḥājj Amīn al-Ḥusaynī for his anti-Zionist struggle in Palestine, which continued Nazi aspirations.[203] It was Salāma Mūsā who condemned the popular racist fanaticism that peaked in German rule, which is the reason that he called Hitler and Chamberlain the two politicians who were the most distant from the European spirit.[204]

We find a similar wave of xenophobia and ethnic nationalism on the northern shore of the Mediterranean. Turkish rule was not unanimously condemned by the Orthodox clergy. For example, Patriarch Anthimos of Jerusalem (1717–1808) praised the Ottoman Empire as being a divine bulwark against the heresies of the West.[205] The image of the West further deteriorated among the population of Southeast European when Western powers became more powerful in the Eastern Mediterranean which, as we have seen, was accompanied by no less power-

---

197 Arsuzi-Elamir, *Arabischer Nationalismus*, op. cit., 241.

198 Barqāwī, *muḥāwala*, op. cit., 83.

199 Karsh, *Islamic Imperialism*, op. cit., 129–30.

200 Cf. Fouad Ajami, *The Dream Palace of the Arabs: A Generation's Odyssey* (New York: Vintage Books, 1999), 62; Kreutz, *Das Ende des levantinischen Zeitalters*, op. cit., fn. 745: 376.

201 Jeffrey Herf, *Nazi Propaganda for the Arab World* (New Haven/Conn.: Yale University Press, 2009), 134.

202 Ibid., 142.

203 Ibid., 243–4.

204 Mūsā, *mā hiya an-nahḍa?*, op. cit., 133.

205 Clogg, "The ‚Dhidhaskalia Patriki'," op. cit., 95–6, cf. ibid. 102.

ful intellectual challenges, to which the Greek Church reacted by protecting their own tradition by means of Aristotelian philosophy.[206]

When freedom of religion was about to be established in post-revolutionary Greece, the Greek bishops unanimously voted for the creation of a holy synod inaugurated by Otto, then king of Greece. This caused a stir among the rank and file, and a monk from Mt. Athos by the name of Prokopios openly called for resistance, arguing that the country would lose its Orthodox imprint and turn either Catholic or Protestant. In this heated atmosphere, the Russian envoy encouraged the Greek bishops to resist a synod, which was finally picked up on by the press. As Georg Ludwig von Maurer reported, the entire public debate had taken a course that was in direct opposition to everything that had been said and proclaimed during the war of freedom and the arrival of King Otto.[207] Instead, the government intended to strengthen the Church's position after it had for centuries been endangered under Ottoman rule.[208]

The Orthodox Church departed even more from the West when it underwent a shift from abstract theology to a living spirituality. Bishop Theophilos of Campania (d. 1795), a student of Voulgaris, spoke of the envisioning of the divine light as being the content of the beatitude of the angels and the just. The Greek theologian Athanasios Parios (1721–1813) even went so far as to defend the Ottoman Empire as being created by God and as being the legitimate heir to the Eastern Roman Empire. He praised it for being a barrier to the Occidentals and a means to salvation for "us Orientals." According to Makrides (2009), there are "astonishing parallels" between Orthodox and Islamic anti-Western policy, and still a "pronounced anti-Occidentalism" exists within Orthodoxy today. To what extent this can be traced back to the teachings of Palamas and their anti-intellectual strain is disputed.[209] In any case, as Beck (1994) has written, Palamism and the opposition to Latin dogmas became the "antipodes" of the Western Renaissance.[210]

Even a reformist such as Evgenios Voulgaris defended the teachings of Palamas against Western theologians such as the Protestant theologian Pierre Leclerc.[211] Likewise, the eighteenth-century Greek Patriarchate attempted to ward

---

206 Podskalsky, *Griechische Theologie*, op. cit., 69.

207 Von Maurer, *Das griechische Volk*, Zweiter Band [2nd vol.], op. cit., 154–8.

208 Ibid., 160.

209 Kreutz, *Zwischen Religion und Politik*, op. cit., 30–1; Podskalsky, *Griechische Theologie*, op. cit., 38–39, 41–2, 44, 60, fn. 1513: 361; Makrides, "Orthodoxes Ost- und Südosteuropa," op. cit., 206, 215.

210 Beck, *Das byzantinische Jahrtausend*, op. cit., 309.

211 Podskalsky, *Griechische Theologie*, op. cit., 38–39, 42.

off Western influence by creating an index of forbidden books.[212] It is little wonder that the concept of "historical pseudomorphosis," which was coined by Oswald Spengler in order to denote the superimposition of an older foreign culture on a younger and native one, thereby hampering the native one in its growth, found an Orthodox echo. Introduced by the Russian-Orthodox theologian Georges Florovsky, the term "pseudomorphosis" could be easily applied to the Western influence on Orthodox culture, where it was rejected as something inauthentic.[213] This anti-Western stance is still alive today.[214]

Similarly, the community of believers in the Russian Orthodox Church is based on the idea of a perfect society embracing among its highest values a unity in God, which encourages the tendency to meet pluralism and individual freedom with skepticism. This might constitute an obstacle to the idea that the rule of law should be based on individual rights and independent institutions, and contrasts with developments in Western Europe.[215] The underlying idea of political freedom is just as little realized in the Islamic context, where egalitarianism is held in much higher esteem. Yet, historically, this did not include women, slaves or infidels – something that reverberates in the Middle East of today.[216]

An anti-Western attitude is also prevalent in the ideas of Ionos Dragoumis (1878–1920), our Greek nationalist of the first half of the twentieth century, who argued that any nation that glued people together created a bond that could not be broken.[217] From this angle, Hellenism, like every nationalism, is to be directed "against the tide of cosmopolitanism," and therefore against Eu-

212 Heyer, *Die Orientalische Frage*, op. cit., 6–7; cf. Wiggers, *Geschichte der Evangelischen Mission* [1845], vol. 1/2: 233. On Pinkerton v.. William Jowett, *Christian Researches in the Mediterranean*, op. cit., 363–5.

213 Buchenau, *Auf russischen Spuren*, op. cit., 364. On the term "historical pseudomorphosis" v. Oswald Spengler, *Der Untergang des Abendlandes: Umrisse einer Morphologie der Weltgeschichte* (Munich: C.H. Beck, 1990), 784, 800, 838.

214 Makrides, "Orthodoxes Ost- und Südosteuropa," op. cit., 206; idem, "Orthodox Anti-Westernism Today: A Hindrance to European Integration?" in *International journal for the Study of the Christian Church*, vol. 9 (3) (2009): 209–224, *passim*.

215 Tobias Traut, "Der Staat im Denken der Russisch-Orthodoxen Kirche: Platz für Demokratie?", in *Religion in Diktatur und Demokratie: Zur Bedeutung religiöser Werte, Praktiken und Institutionen in politischen Transformationsprozessen*, ed. Simon Wolfgang Fuchs and Stephanie Garling (Münster und Berlin: Lit, 2011): 59–78, here 65–6, 73. Fukuyama, *Political Order and Political Decay*, op. cit., 11; cf. Rhonheimer, op. cit, 349–50.

216 Cook, *Ancient Religions*, op. cit., 321, cf. ibid., 248, 334, 390.

217 "[...] Καί τό ἔθνος εἶναι ἕνας δεσμός, καί δέν μποροῦμε μεῖς νά τόν σπάσωμε." Dragoumis, Ὁ Ἑλληνισμός μου καί οἱ Ἕλληνες, op. cit., 91.

rope, which is thought to be characterized by cosmopolitanism.[218] Neither ancient Greece nor the Eastern Roman Empire can provide orientation for the future of Greece, but a united Greece within the borders of the "Greek race" is the goal – it is "where we go."[219] Dragoumis also expressed anti-American resentment when he passed on a story told by a friend to the effect that Americans dump a book when they do not find it immediately useful.[220]

Dragoumis drew heavily on the ideas of, for example, Maurice Barrès, a proponent of a xenophobic ethnic nationalism who also enjoyed some popularity in the Arab world. In Alexandria, it was Christian Lebanese nationalists such as Michel Chiha (who was born to a Chaldean Catholic family).[221] Chiha became acquainted with Hector Klat (born to a Greek Orthodox family), a poet who after the war would become one of the most influential proponents of the so-called Phoenician ideal. Klat, in turn, was in personal contact with Maurice Barrès, who had travelled to the Middle East where he met several intellectuals on both sides of the Levantine shore, including Ionos Dragoumis, whom he left impressed.[222]

The xenophobia and ethnocentrism of Barrès reverberates in Dragoumis' ideal of "archaism" and his attack on "xenolatreia" (lit. "foreignism"), an ideology that viewed the Franks, i. e., the Europeans, as role models for Greece, which he dismissed as a terrible idea. He also opposed a modern secular constitution for the Greek polity because the Greeks could draw on more authentic resources, such as the reign of Theseus, Athenian democracy, and the Spartan regime. In-

218 "Ἐναντίον στό ρεῦμα τοῦ κοσμοπολιτισμοῦ πηγαίνει τό ρεῦμα τοῦ Ἑλληνισμοῦ καί κάθε ἐθνισμοῦ. Ὁ πολιτισμός τῆς Εὐρώπης ὁ τωρινός πηγαίνει κατά τόν κοσμοπολιτισμό. Οἱ περοσότεροι ἀφήνονται στό ρεῦμα αὐτό. [...]" Ibid., 87.

219 "[...] Οὔτε ἡ Ἑλλάδα ἡ ἀρχαία εἶναι τό πρότυπο τῆς τωρινῆς οὔτε ἡ βυζαντινή αὐτοκρατορία, μά μιά Ἑλλάδα που νἄχη σύνορα τά σύνορα τῆς ἑλληνικῆς φυλῆς. Ἐκεῖ πηγαίνομε." Ibid., 120.

220 "Κάποιος μέ διηγήθηκε πως οἱ Βόρειοι Ἀμερικάνοι μόλις [kaum dass] διαβάσουν ἕνα βιβλίο θέλουν νά ἐφαρμόσουν τίς ἰδέες που περιέχει. Προσπαθοῦν καί δοκιμάζουν. Καί ἄν ἐφαρμόζωνται οἱ ἰδέες αὐτές καί μποροῦν νά γίνουν πράξη, τότε θεωροῦν τό βιβλίο χρήσιμο, εἰδεμή τό πετοῦν στά σκουπίδια." Ibid., 133.

221 Xydis, "Modern Greek Nationalism," op. cit., 245; Zelepos, *Die Ethnisierung griechischer Identität*, op. cit., 215 Fn. 307; cf. Kreutz, *Das Ende des levantinischen Zeitalters*, op. cit., 365 fn. 629. Dragoumis, *Ὁ Ἑλληνισμός μου καί οἱ Ἕλληνες*, op. cit., chap. Γ, β Alexandria 1905, § 4: 62. Kaufman, *Reviving Phoenicia*, op. cit., 159.

222 Zelepos, *Die Ethnisierung griechischer Identität*, op. cit., 215 fn. 307; Kaufman "Phoenicianism", op. cit., 183–4; idem, *Reviving Phoenicia*, op. cit., 63–5.

stead, everything European, he complained, was preferred to one's own culture.[223]

In this respect, he goes on to argue, it is not the Russians, Bulgarians, Romanians, Austrians, Italians or Turks who press hard on the Greeks, but the *idées modernes*, i.e., contemporary civilization "with its masonry, its philanthropy, its solidarity, its parliamentarianism and its egalitarianism, which make all humans equal to minors," since it erases differences and grinds the corners of identity round, "like the stones of a river where the water makes them pebble stones." He calls on the Greeks to stop "this invasion of contemporary culture" and to stifle its destructive impact before they are downgraded to the status of mere "herd animals." Dragoumis makes this a question of life and death.[224]

The Greeks, he argues, are not a new race, but carry a burden of historical memory.[225] Against this backdrop, he wonders why they imitate European – i.e.,., Western European – habits and ideas, and he praises the alleged superiority of traditional Greek culture: "In our blood we have a culture better than anyone else's."[226] From this perspective, there is no civilization more advanced than Greek civilization, and nor will there be. Since the Greeks were the first humans, they represent humanity perfectly, and no other people can hold a candle to

---

223 "Ὄχι, δέν μᾶς ἔβλαψε ἡ ἀρχαιοπρέπεια, ὁ ἀρχαϊσμός. Μάς ἔβλαψε ἡ ξενολατρεία, ὁ ξενισμός. Δυό πράγματα θελήσαμε πολύ νά μιμηθοῦμε γιά νά γίνωμε ἄνθρωποι· α) τούς ἀρχαίους β) τούς ξένους, τούς Φράγκους, τούς Εὐρωπαίους. Οἱ ἀρχαίοι δέν μᾶς ἔβλαψαν. Μᾶς ἔβλαψαν φοβερά οἱ ξένοι. Καλλίτερα θά ἦταν, χίλιες φορές, ἄν, ἀντί νά διαλέγαμε τό σύνταγμα γιά πολίτευμά μας, διαλέγαμε τήν βασιλεία τοῦ Θησέα, τοῦ Κόδρου, ἤ τήν ἀθηναϊκή δημοκρατία, ἤ τό σπαρτιατικό πολίτευμα. Μᾶς ζάλισε πάρα πολύ ὁ εὐρωπαϊκός πολιτισμός." Dragoumis, *Ὁ Ἑλληνισμός μου καί οἱ Ἕλληνες*, op. cit., 77.

224 "Δέν εἶναι οἱ Ρῶσσοι, οἱ Βούλγαροι, οἱ Ρουμάνοι, οἱ Αὐστριακοί, οἱ Ἰταλοί, οἱ Τοῦρκοι, που μᾶς ἐπιβουλεύονται. [...] Μᾶς ἐπιβουλεύονται οἱ *idées modernes*, ὁ σύγχρονος πολιτισμός μέ τόν μασονισμό του, μέ τήν φιλανθρωπία του, μέ τήν ἀλληλοβοήθειά του, μέ τόν κοινοβουλευτισμό του, μέ τό ἰσοπεδωμά [του, που κάνει ὅλους τούς ἀνθρώπους ἴσους μέ τούς μικρότερους, ὅλος αὐτός ὁ σύγχρονος πολιτισμός [...] που σβήνει τίς διαφορές, κόβει τίς γωνίες καί μᾶς στρογγυλεύει σάν τίς πέτρες τοῦ ποταμοῦ, που τίς κάνει χαλίκια τό νερό. Αὐτοῦ τοῦ σύγχρονου πολιτισμοῦ ἡ εἰσβολή [...] μᾶς σαπίζει καί μᾶς μολῦνει [...]. Αὐτός εἶναι ὁ πολιτισμός που μᾶς κάνει ζῶα κοπαδιοῦ, καθώς λέγει ὁ Nietzsche. Καί αὐτοῦ τοῦ πολιτισμοῦ τήν ἐπίδραση πρέπει νά πολεμήσωμε, ἄν θέλωμε νά ζήσωμε." Ibid., 81.

225 "Δέν εἴμαστε νέα φυλή· στό αἷμα μας εἶναι μέσα καί τό φαρμακώνει ἡ ἱστορία μας· καί εἴμαστε κουρασμένοι ἀπό τό βάρος τοῦ μνημονικοῦ μας (τῆς μνήμης τῆς ἱστορίας)." Ibid., 95.

226 "Τί μιμοῦνται οἱ Ἕλληνες τίς εὐρωπαϊκές συνήθειες καί ιδέες! Μεῖς ἔχομε στό αἷμα μας μέσα ἕναν πολιτισμό καλλίτερο ἀπό κάθε ἄλλον. [...]" Ibid., 119. "Εἶναι ἀνώτερη ἀπό πολλές ἄλλες ἡ ἑλληνική φυλή, ἀφοῦ καταφέρνει τόσους αἰῶνες τώρα νά κυριαρχῇ ἐπάνω στίς ἄλλες τριγυρινές φυλές. Ὅταν 14 Ἕλληνες δεσποτάδες κυβερνοῦν χιλιάδες Σύρους καί οἱ Σύροι αὐτοί παραπονοῦνται πως εἶναι εἵλωτες τῶν δεσποτάδων, τί ἄλλο θέλετε γιά νά ἀποδειχτῇ πως οἱ Ἕλληνες ἀξίζουν κάτι ὡς κυβερνῆτες." Ibid., 133–4.

them. Other cultures have done nothing but create anew what the Greeks had already created, so that it does not make sense for the Greeks to adopt their culture.[227]

Dragoumis was not alone in his feelings of supremacism regarding other nations. Imbued with an anti-modernist trait, this attitude was compatible with anti-Semitic thoughts, not only in Greece, but also, for example, in Orthodox Serbia, too. The Serbian Orthodox theologian Nikolay Velimirovic depicted the Jews at the beginning of the twentieth century as the perfect others vis-à-vis the Serbs, and as agents of a secular and therefore sinister modernity.[228] This was also the background for the attempt made by Nazi Germany to initiate a special program aimed at finding common ground with Orthodox theologians.[229]

3.4 After Disillusionment

Despite the authoritarian and xenophobic traits in Dragoumis' thinking, he was a keen observer of the political situation. His 1904 remarks on the dysfunctionality of a situation in which Greece is run by tyrants is quite lucid and straight to the point.[230] For example, he criticizes his compatriots heavily for being "oligarchic rather than democratic,"[231] and deplores a complete lack of accountability in political matters – something that still sounds all too familiar in present-day Greece. This also holds true for the Greeks who are part of the diaspora, who would reportedly say, "You, those in Greece, should do this to save us," while the homeland Greeks would say, "You, out there, you have great power; why

227 "Οἱ Ἕλληνες βλέποντας τόν κόσμο (δηλαδή τόν ἄνθρωπο) ἀπό μιά θέση δική τους τόν παράστησαν κάπως, ἀλλά τέλεια. Οἱ Εὐρωπαῖοι βλέποντάς τον ἀπό μιά ἄλλη θέση, λίγο παρακεῖ τόν παράστησαν καί αὐτοί κάπως, ἴσως κι αὐτοί τέλεια, μά ὄχι διαφορετικά τέλεια ἀπό τοῦς Ἕλληνες. [...] Δηλαδή κανένας πολιτισμός δέν ἔφτασε ψηλότερα ἀπό τόν ἑλληνικό οὔτε θά φτάση. Οἱ Ἕλληνες ἦταν οἱ πρῶτοι ἄνθρωποι καί οἱ τέλειοι. Αὐτοί πρώτοι παράστησαν τόν ἄνθρωποι τέλεια. Ἄν παραδεχτοῦμε ὅτι ὁ ἑλληνικός ἔχει χρῶμα ἄσπρο καί κόκκινο καί μαῦρο καί μόρικο, οἱ ἄλλοι ἔχουν μερικά χρώματα ἤ συνδυασμούς χρωμάτων διαφορετικούς λιγάκι. [...] Οἱ Ἕλληνες ἀπό ἐξαιρετική ἀγάπη γιά τόν ἄνθρωπο ἔπλασαν τόν ἄνθρωπο. Καί ὕστερα οἱ ἄλλοι πολιτισμοί δέν κάνουν τίποτε ἄλλο, παρά νά τόν ξαναπλάθουν. Μά θά μοῦ πῆ κανείς· δέν εἶναι περιττό νά ξανακάνουν οἱ ἄνθρωποιτά ἴδια πράματα; [...]" Ibid., 145–6.

228 Buchenau, *Auf russischen Spuren*, op. cit., 458, 460.

229 Ibid., 364–5.

230 "Οἱ δημοκρατίες εἶναι ἑλληνικό πολίτευμα, ὀλιγαρχικές ὅμως δημοκρατίες. Κάθε πολιτεία ἑλληνική, κάθε ἀποικία τό ἔχει καί τό εἶχε. Καί μόνον οἱ τύραννοι ἔκαμαν πολιτική δουλειά ὅπως πρέπει." Dragoumis, *Ὁ Ἑλληνισμός μου καί οἱ Ἕλληνες*, op. cit., chap. B, § 13: 37.

231 "Οἱ Ἕλληνες εἴμεθα ὀλιγαρχικοί καί ὄχι δημοκρατικοί καί ὅμως συχνά χανόμαστε ἀπό τά λόγια καί δέν τό καταλαβαίνομε πως δέν εἴμαστε δημοκρατικοί." Ibid., 73.

don't you fight this or that and not save yourselves?" Dragoumis concludes that everyone pins the blame for the lack of initiative on someone else.[232]

Georg Ludwig von Maurer already described in the nineteenth century the situation in Greece, where the captains (*ocaks*) and *kocabasis* ruled civil life, maintained their own courts, and were at loggerheads with each other. While serving their Ottoman occupiers and oppressing the Greek population, they were paranoid regarding their environment and feared its hostility. This might help explain the ill state of the Peloponnese.[233] Namely, in the Greek case, as Fukuyama (2014) has called this a "modernization without progress."[234] Greece, however, never managed to reform its public sector as this was the case, for example, in Germany, Great Britain and the US.[235] In this regard, Greece "fits a pattern more characteristic of many contemporary developing societies elsewhere in the Balkans, the Middle East, and Africa."[236]

The Orthodox Church in Greece is officially recognized today as the "prevalent religion" of the country, and is generally understood as a state religion.[237] This has a lasting effect on its self-image as being part of the Orthodox, non-Western world. This, in the words of Makrides (2000), adds to Greece's being a "torn country," insofar as "its political and economic elites are pro-Western, but its history, culture, religion and tradition are essentially non-Western."[238]

A similar pattern emerged in the Arab-Islamic world. Starting out with the influence of French law in the Ottoman Empire, the Arab-Islamic world seemed at the beginning to be embarking on a path to modernity that was based on individual rights and the rule of law. While under British occupation (1882–1936), Egypt saw the civil rights and liberties of Egyptians being recognized officially,

---

232 "Οἱ ἔξω Ἕλληνες λέν: "Σεῖς, αὐτοῦ στήν Ἑλλάδα, πρέπει νά κάμετε τοῦτο γιά νά μᾶς σῶστε". / Οἱ μέσα Ἕλληνες λέν: "Σεῖς, αὐτοῦ ἔξω, ἔχετε μεγάλη δύναμη· γιατί δέν πολεμᾶτε τοῦτο ἤ ἐκεῖνο καί δέν σώζεσθε;" / Καί ἔτσι ὁ καθένας ρίχνει τήν εὐθύνη καί τήν πρωτοβουλία τῆς δουλειᾶς, τῆς ἐθνικῆς ἐργασίας, στόν ἄλλο. Καί τίποτε δέ γίνεται." Ibid., chap. B, § 4: 41.

233 Von Maurer, *Das griechische Volk*, Erster Band [1st vol.], op. cit., 46–8.

234 Fukuyama, *Political Order and Political Decay*, op. cit., 117.

235 Ibid., 122.

236 Ibid., 99.

237 Makrides, "Orthodoxes Ost- und Südosteuropa," op. cit., 213, 215.

238 From the *Encyclopedia of Greece and the Hellenic tradition*, lemma "Anti-westernism." Huntington: "Greece is not part of Western civilization, but it was the home of Classical civilization which was an important source of Western civilization. ... Unlike Serbs, Romanians, or Bulgarians, their history has been intimately entwined with that of the West. Yet Greece is also an anomaly, the Orthodox outsider in Western organizations." Huntington, *The Clash of Civilizations*, op. cit., 162.

and suffrage being granted in 1883.[239] Of course, Western powers never played an entirely progressive role in the formation of the modern Middle East, and we must keep in mind the bombing of Alexandria as well as other incidents so as not to fall into the trap of excusing colonialism.[240] However, this is only part of a long history of violence in the Middle East, and the Ottoman Empire has its own record of violence and oppression. In any case, the spirit of progress of the nineteenth and early twentieth century would not last too long.

While in Egypt all religious courts and the minority parliaments (*majālis millīya*) of the Jews and Christians were abolished in 1956, and a family law that was not *sharia*-based was introduced in 1979, the country suffered a setback in 1980 when the *sharia* was upgraded from being one source of the constitution to its main one. It was the first step to the revision of family law that took place in 1985.[241] We can see a similar development in Turkey, where, with the establishment of the republic, first the sultanate was abolished in 1923, and then also the caliphate in 1924.[242] At the same time, all existing 479 madrasas were closed down and religious education at state schools terminated.[243] A process of re-Islamicization began in 1961 at the latest, when the Presidency of Religious Affairs (*Diyanet İşleri Reisliği*) became a public law institution. Since then, Islam has been the *de facto* state religion.[244] Yet, the blueprint of modernization still comes from the West.[245]

In this regard, Ghalioun (2007) might be correct when he asserts that the Arabic civilization of today lives practically in a state of mental colonization while condemning the free and active personality. All the great civilizations, he writes, have had the same experience when challenged by the achievements of a more successful civilization.[246] However, he rejects the idea that the contemporary crisis of identity is a mere outcome of foreign cultural hegemony. Rather, when a civilization forfeits its creative energies, then it is soon dominated by another emergent civilization, and hides itself away. It degenerates from a creative

---

239 Anderson, "Law Reform in Egypt," op. cit., 158–9.
240 Mūsā, *mā hiya an-nahḍa?*, op. cit., 116.
241 Thielmann, *Naṣr Ḥāmid Abū Zaid*, op. cit., 68–9, 71.
242 Karl Binswanger, "Türkei," in *Der Islam in der Gegenwart: Entwicklung und Ausbreitung. Staat, Politik und Recht, Kultur und Religion*, ed. Werner Ende and Udo Steinbach (Munich: C.H. Beck, 1991): 212–20, here 213.
243 Ibid.
244 Ibid., 217–8.
245 Cook, *Ancient Religions*, op. cit., 440.
246 Ghalioun, *naqd as-siyāsa*, op. cit., 253.

to a mere consuming civilization, before fragmenting into smaller national and linguistic entities.[247]

In such a situation, an emerging feeling of powerlessness can give rise to a radical notion of religion, which seeks to intrude into politics. According to Ziedan (2010), religion and politics operate in principle within their own orbit and stay on their own course.[248] Since politics is based on society, and religion on the individual, there is nonetheless an "elliptic overlay" between the two. Therefore, politics can control society through the individual; and, conversely, the individual can establish his or her own specific religious certitude through the stock of divinity provided by society.[249] On the other hand, although everyday life in the Muslim world does not always comply with the tenets of faith, there is no dualism between the spiritual and the profane.[250]

At the same time, the project of reforming Islam has stalled. The project has not yielded results powerful enough to advance Muslim societies so that they embrace an increase in individual freedom, political participation, devolution of power, and economic prosperity. Instead of introducing reform worthy of its name, we see, as Lazarus-Yafeh (1992) has put it, "an immense modernist-apologetic literature claiming that the roots of most Western values, sciences and technology are explicitly mentioned in the Quran."[251] This is in fact ahistorical and misleading since it is no more than a backward projection.

Similarly, a contemporary reformist such as Burhan Ghalioun praises speculative theology (*ʿilm al-kalām*) and Muslim philosophy for having shaped Western thought more than Greek philosophy has ever done. According to this view, Averroes was pivotal in medieval Western thinking, since he emphasized the significance of reason for the contemplation of the divine, with reason never having been a major issue before the rise of Christianity and Islam. Ghalioun even goes so far as to say that the religious philosophy of Muslim philosophers such as al-Ghazālī and al-Fārābī was nothing less than the foundation which Western medieval thinking was built.[252]

247 Ibid., 248.

248 Youssef Ziedan [Yūsif Zīdān], [Arab.] *al-lāhūt al-ʿarabī wa-uṣūl al-ʿunf ad-dīnī* (Cairo: Dar ash-Shuruq, 2010), 212.

249 Ibid., 213.

250 Tilman Nagel, "Gab es in der islamischen Geschichte Ansätze zu einer Säkularisierung?", in *Studien zur Geschichte und Kultur des Vorderen Orients: Festschrift für Bertold Spuler zum siebzigsten Geburtstag*, ed. Hans R. Roemer and Albrecht Noth (Leiden: Brill 1981): 275–88, here 275–7.

251 Lazarus-Yafeh, "Die islamische Reaktion," op. cit., 218.

252 Ghalioun, *naqd as-siyāsa*, op. cit., 250–1.

Irrespective of its accuracy, this line of thinking is typical of many modern Muslim reformist thinkers, who find intellectual satisfaction in the idea that the West owes its cultural foundations to the Arab and Islamic influence. They often believe that science no longer found a home in the Islamic world when Muslims turned away from their religion and therefore paved the way for the progress of the West. Only a return to the true religion would make their civilization blossom again. From this angle, science is not an open system to which anyone can contribute, but something that the West has taken away from the East and has only developed it further. By thinking in such a way, Ghalioun isolates scientific progress from its social and historical conditions.[253]

This approach has been sharply criticized by Strohmaier (2003), who argues that highlighting the Muslim contribution to Western culture usually includes an overestimation of the influence that Arabic-Islamic texts had on cultural processes in Western Europe – with an emphasis on "Western," since the Eastern part of the continent is to be considered separately.[254] Deviant thinkers such as Abū Zayd (1992), who called for Muslims to begin the stage of self-liberation, not only from the authority of texts alone, but from every authority that restrains the human in his or her journey in the world, remain largely unheard.[255]

---

253 Barqāwī, *muḥāwala*, op. cit., 52.

254 Gotthard Strohmaier, "Was Europa dem Islam verdankt," in idem, *Hellas im Islam: Interdisziplinäre Studien zur Ikonographie, Wissenschaft und Religionsgeschichte* (Wiesbaden: Harrassowitz, 2003): 1–27, here 25–6.

255 Abū Zayd, *al-Imām ash-Shāfiʿī*, op. cit., 110.

4 Concluding Thoughts

There are no irrevocable triumphs of consciousness over its downfalls, as Hans Blumenberg (2001) once put it. Why the ideas of the Enlightenment enshrined in the cultural revivalisms of the Eastern Mediterranean were finally watered down and failed to prevail is key to understanding the modern Eastern Mediterranean, including the Middle East.[1] Sources assume that individual freedom was never really fleshed out and did not find its way into the modern rule of law, good governance and democratic institutions. We should also not be blind to the fact that there is still a difference between the predominantly Christian Orthodox countries and those of the Muslim world.

However, modern-day postcolonialism, which is prevalent at Western universities, shows little interest in Enlightenment processes outside Europe, since it aims to show that the cultures of the world might be entangled, but it does so only to the effect that Europe is indebted to the Muslim Middle East, while regarding the Western influence on the Muslim world as despicable and destructive. As a consequence, the cultural resources of Islam contain everything necessary to build a modern society. Steinberg (2008) argues that the ideology of postcolonialism "has replaced research with systematic biases that select favored 'victims' and rejected 'oppressors', and empirical methodology based on testable hypotheses with political formulae and incantations."[2]

Detecting and rejecting Orientalism and Eurocentrism in all their guises have therefore become preferable in postcolonialism to dealing with facts and sources. It is for this reason that contemporary publications on the modern Middle East downplay or even ignore the Nahḍa, since it was heavily influenced by French thought. A 2014 book on the history of Egypt omits the Nahḍa from history completely.[3] The ideological motivation behind this kind of publication can be called Islamic supremacism, but it is not confined to Muslim apologists of Islam; it has become the bedrock of Middle Eastern studies and adjacent fields.

---

1 Hans Blumenberg, "Wirklichkeitsbegriff und Wirkungspotential des Mythos," in *Terror und Spiel*, ed. Manfred Fuhrmann (Munich: W. Fink, 1971), 11–66, repr. in Hans Blumenberg, *Ästhetische und metaphorologische Schriften*, ed. Anselm Haverkamp (Frankfurt/Main: Suhrkamp, 2001): 327–405, here 345.

2 Gerald M. Steinberg, "Postcolonial Theory and the Ideology of Peace Studies," in *Postcolonial Theory and the Arab-Israel Conflict*, ed. Philip Carl Salzman and Donna Robinson Divine (London and New York: Routledge, 2008): 109–19, here 118.

3 Johanna Pink, *Geschichte Ägyptens: Von der Spätantike bis zur Gegenwart* (Munich: C.H. Beck, 2014). This is all the more confusing against the background that the book is advertised as essential for understanding why Egypt today is a leading intellectual, cultural and political force.

https://doi.org/10.1515/9783110634006-006

Postcolonialism in Middle Eastern studies at Western universities is largely identical to Edward Said's school of thought, which is based on two assumptions. First, that Islam explains nothing. And, second, that attributing any greater problem that might occur in the Islamic world to Islam must be the result of Western (Eurocentric) projection. As Said put it himself, he tried "to show that European culture gained in strength and identity by setting itself off against the Orient as a sort of surrogate and even underground self."[4] Said modified his views in his 1993 book *Culture and Imperialism*, where he conceded that his remarks had not intended "to denigrate the accomplishments of many Western scholars, historians, artists, philosophers, musicians, and missionaries, whose corporate and individual efforts in making known the world beyond Europe are a stunning achievement."[5]

However, Said criticized scholars for producing findings that ran counter to "native Arab or Islamic nationalism,"[6] which by extrapolation suggests that Middle Eastern studies are expected to popularize nationalist narratives for their own sake. This also runs counter to another popular assumption of postcolonialism: namely, that national identities are "socially constructed" and do not reflect hard facts. Fukuyama is correct to argue that this might be possible, but he asks the question of who constructs such an identity, and whether such a construction of identity is top-down or bottom-up.[7] In any case, it seems that, when it comes to the Islamic world, postcolonialism is happy to nullify its own assumptions.

As Lewis (2012) has noted, the "situation in Near East studies is a great detriment to the state of scholarship in the field. This is part of a general change, a political correctness, in which Islam now enjoys a level of immunity from comment or criticism in the Western world that Christianity has lost and Judaism has never had."[8] Similarly, Osterhammel (1998) has argued that Said's *Orientalism* paved the way for a kind of criticism that comes "with an air of unmasking and one-sided exaggeration."[9]

When Said argued that it was a "standard misrepresentation" to believe that "exclusively Western ideas of freedom led the fight against colonial rule, which mischievously overlooks the reserves in Indian and Arab culture that always re-

---

4 Said, *Orientalism*, op. cit., 3.

5 Said, *Culture and Imperialism*, op. cit., 235.

6 Ibid., 315.

7 Fukuyama, *Political Order and Political Decay*, op. cit., 191.

8 Bernard Lewis (and Buntzie E. Churchill), *Notes on a Century: Reflections of a Middle East Historian* (New York: Viking, 2012), 271.

9 Osterhammel, *Die Entzauberung Asiens*, op. cit., 18.

sisted imperialism,"[10] then he is correct in his emphasis on "exclusively." However, whether we like it or not, the Western impact is a fact and was not always negative,[11] since the blueprints for modernization came, in one way or another, from the West.[12]

In any case, Mediterranean perspectives and interdisciplinary approaches are a promising way to understand the modern Middle East and Southeast Europe as well. We need to pay more attention to cultural entanglements and overlapping civilizations, so as to be able to examine the dynamics of state building and societal change. Mapped along the lines of political reformist thinking and corresponding events, comparative approaches focusing on different, but overlapping, cultures and regions have the potential to yield new insights into related issues. This is the key to understanding the modern societies of the Eastern Mediterranean.

10 Said, *Culture and Imperialism*, op. cit., 241–2.

11 Ibid., 263–4. On Wollstonecraft v. Kreutz, *Zwischen Religion und Politik*, op. cit., 197.

12 Cook, *Ancient Religions*, op. cit., 4, 440.

Timeline

639	Islamic conquest of Egypt.
691–2	Beginning of the schism between the Eastern and the Western Church, marked by the Council in Trullo.
863	Bardas' re-founding of the University Constantinople, which becomes the center of a renewed Hellenism.
1053	The Great Schism between the Eastern Orthodox Church and the Western Catholic Church.
1270	The last major crusade in the Near East.
1351	The teachings of Gregory Palamas become the official doctrine of the Orthodox Church.
1354	Capture of the Greek theologian Gregory Palamas by the Ottomans.
1368	Canonization of Palamas by the Church.
1439	Exacerbation of the theological split between the Eastern and the Western Church by the Florentine Union.
1584	Publication of Martin Crusius' edited *Turcograecia*.
16th ct.	Intensification of Protestant contacts with Greek theologians.
	Program of modernization in Persia under Shāh ʿAbbās (r. 1588–1626).
1600	Britain's founding of the East India Company.
17th ct.	Increasing presence by the beginning of the century of Western powers in the Eastern Mediterranean.
	Establishment of commercial ties by the middle of the century between Greeks and Western Europe.
1606	The first recognition by the Ottoman Empire of a European power as a negotiating partner on equal terms in the peace treaty of Zsitvatorok.
1622	Theofilos Koridalleas (1570–1646) becomes head of the Patriarchal Academy of Constantinople.
1623	Persia breaks the Portuguese maritime trade monopoly.
1648	Peace of Westphalia.
1683	Printing in Leiden (Netherlands) of Patriarch Kyrillos Loukaris' translation of the New Testament into contemporary Greek.
1697	Publication of Barthélemi d'Herbelot's *Bibliothèque orientale*.
1711–6	Beginning of reign of Phanariots in Moldavia and Wallachian.
1718	Treaty of Passarowitz between Habsburg and the Ottoman Empire.
1719	Founding of the second Oriental Company in Vienna.

https://doi.org/10.1515/9783110634006-007

1723	Accusation of heresy leveled by the Patriarchal Synod against Methodios Anthrakitis of Epirus.
1727	Beginning of the Archimandrite Theokletos Polyeidis' journey through Germany.
1749/50	Founding of the Athonias academy at the monastery of Vatopedi by Patriarch Kyrillos V.
1751	Publication of a contemporary Greek translation of the historical account of Agathangelos, a supposed Armenian monk of the fourth century.
1764	Publication of Voulgaris' translation of *Historical Critical Essay on the Quarrels within the Church of Poland.*
1765, ca.	Appearance of Evgenios Voulgaris in the city of Leipzig.
1769–70	Uprisings against the Ottomans on the Greek Peloponnese, encouraged by the Russian empress Catherine the Great.
1771	Evgenios Voulgaris heads for Russia. Athanasios Parios becomes head of the Athos academy.
1772, ca.	Publication of Voulgaris' *Thoughts on the Current Crisis of the Ottoman State.*
1772–3	Publication of August Ludwig Schlözer's *Universal History.*
1773	Diderot resides at Catherine the Great's court.
1774	Russia and the Ottoman Empire make peace in the treaty of Küçük Kaynarca, which Russia interprets as legitimizing a Russian protectorate over the Christians in the Balkans.
1775	Eugenios Voulgaris becomes archbishop of the newly founded diocese of Kherson.
1783	Catherine the Great annexes the Crimea.
1785	Publication of Antonios Moschopoulos' translation of Wolff's *Logik.*
1791	Publication of Christodoulos Pamplekis' *Trophies of the Orthodox Faith.*
1793	Condemnation by the Eastern Church of the writings of Christodoulos Pamplekis.
1797	Publication of Rigas Ferraios Velestinlis' book on *Greek Democracy* (Η ελληνική δημοκρατία). Velestinlis is taken into custody in Trieste. Massacre in Smyrna peaks in the killing of some 2,000 Greeks. France annexes the Ionian Islands.
1798	Anonymous publication of Athanasios Parios' *Christian Apology.* Napoleon conquers Egypt. Velestinlis prepares for the independence of Greece while in Vienna; he is executed in Belgrade.

1799–1806	Ionian Islands fall under Russian-Turkish control.
1801	Withdrawal of France from Egypt after defeat by Anglo-Ottoman troops.
1807	Establishment of the Ionian Academy.
1809–11	Lord Byron travels to the Balkans.
1811–1819	Unrest in Arabia; clamp-down by Muḥammad ʿAlī.
1812–22	Publication of Joseph-François Michaud's account of the crusades (*Histoire des Croisades*).
1814	Founding by Greek merchants of the Society of Friends in preparation for a Greek revolt.
1814/15	Congress of Vienna aims to reorganize Europe after the end of Napoleon's rule.
1816	Egypt's military schools open their gates.
1820–1822	Egypt's Muḥammad ʿAlī conquers Sudan.
1821	Greek revolt against the Ottomans on 25 March.
1822	Inauguration of the national charter, the so-called Constitution of Epidauros.
1822–1826	Egypt's Muḥammad ʿAlī fights the Greeks.
1823	Founding of the Ionian University by the philhellene Lord Guilford.
1824	The Maronite Patriarchate bans both the King James Bible and the Protestant mission for their heresy.
	Death in Messolonghi of Lord Byron while fighting on the Greek side against the Ottomans.
1826	Founding by Egypt's Muḥammad ʿAlī of a hospital and a medical school for the treatment of his troops.
	Elimination in a bloodbath in Constantinople of the Janissaries.
1826–31	Together with students from Egypt, Ṭahṭāwī travels to Paris.
1827	Founding of the Triple Alliance of France, England and Russia, followed by Prussia.
	Publication of Archimandrite Hilarion's translation of the Gospels into contemporary Greek.
1832	Publication of Michel Chevalier's *Système de la Méditerranée.*
1833	French Saint-Simonists travel to Egypt on a campaign to make peace between east and west.
	The Church of Greece declares its independence.
1834	Muḥammad ʿAlī proclaims the independence of Egypt.
1835	Publication in two volumes of Georg Ludwig von Maurer's account of *The Greek People before and after Its Struggle for Independence* (*Das griechische Volk in öffentlicher, kirchlicher und privatrechtlicher Beziehung vor und nach dem Freiheitskampfe bis zum 31. Juli 1834*).

1837	Publication of Thomas Carlyle's *History of the French Revolution.*
1838	Signing of a treaty of commerce between England and the Ottoman Empire.
1839	Proclamation by the Ottoman sultan of the edict of Gülhane for administrative reform (*tanẓīmāt*).
1843	Christian subjects are given equal treatment under civil law due to the *tanẓīmāt* reforms.
1844	Completion by the British and Foreign Bible Society (BFBS) on Corfu of a translation of the Hebrew Bible into contemporary Greek.
1845	Implementation by Egypt's Muḥammad ʿAlī's of a new law, the *qānūn al-muntaḫab.*
1855	The Church of Greece attains the status of a fully fledged autocephaly.
	Publication by the American Bible Society of Elias Riggs' translation of the Gospels into contemporary Greek.
	Publication of Aḥmad Fāris ash-Shidyāq's book *as-sāq ʿalā s-sāq* (*Thigh over Thigh*).
1856	Recognition of the equality of all subjects in the Ottoman *hatt-ı hümayun* decree.
	The Treaty of Paris settles the Crimean War and weakens Russia's role in Southeast Europe.
1858/59	The Maronite Revolt in Lebanon under the leadership of Ṭanyūs Shāhīn creates the Keserwan community.
1860	Upheaval of Christian peasants in Lebanon under the leadership of Yūsif Beg Karam.
1864	Greece drafts a new constitution.
1866	Publication of Francis Fatḥallāh al-Marrāsh's book *ghābat al-ḥaqq*, a dialogue on political freedom.
1867/8	Beginning of the *Meiji* era, which sees Japan reaching out to the West.
1868	Establishment of the *École Khédievale de droit* in Egypt.
1869	Ceremonies to mark the opening of the Suez Canal and the Cairo Opera.
	The reformist Midḥat Pasha becomes provincial governor of Baghdad.
1875	Establishment of mixed courts in Egypt.
1875–8	Orient crisis.
1876	Reforms by Midḥat Pasha lead to the drafting of an Ottoman constitution.
1877	Suspension of the Ottoman constitution by Sultan ʿAbdülḥamīd.

1878	Independence of Serbia, Romania and Bulgaria.
1882–1936	Egypt falls under British occupation.
1883	Official recognition of the civil rights and liberties of Egyptians, and granting of suffrage.
1889	Founding of the Committee of Union and Progress, which aims to reinstate the Ottoman constitution.
1890	Systematic training of lawyers in Cairo.
1898	Publication of Faraḥ Anṭūn's article on "The decline of the Orient."
1899	Publication of Qāsim Amīn's book on the liberation of women (*tahrīr al-mar'a*).
1902	Karl Krumbacher delivers a lecture on the problem of modern Greek literary language ("Das Problem der neugriechischen Schriftsprache").
1903–11	Publication of Ionos Dragoumis' book on *Hellenism and the Greeks* ('Ο 'Ελληνισμός μου καί οἱ "Ελληνες).
1905	Victory of Japan over Russia.
1908	Revolution of the Young Turks and reinstatement of the Ottoman constitution.
	Publication of Sulaymān al-Bustānī's book on the Ottoman constitution, *ʿibra wa-dhikrā*.
1910	Publication of Salāma Mūsā's book on the super-human, *muqaddimat al-subirmān*.
1912	Independence of Albania.
	Publication of Jurjī Zaydān's book *Trip to Europe* (*riḥla ilā Ūrūbā*).
1913	Systematic training of lawyers in Beirut.
1919	Systematic training of lawyers in Damascus.
1921	Publication of Ṭāhā Ḥusayn's treatise on the Athenian *polis* (*niẓām al-Athīniyīn*).
1923	Proclamation of the Turkish Republic.
1924	Abolition by Turkey of the sultanate.
1925	Publication of ʿAlī ʿAbdarrāziq's book on Islam and statecraft (*al-islām wa-uṣūl al-ḥukm*).
	Publication of Ṭāhā Ḥusayn's book *qādat al-fikr* (*The Pioneers of Thought*).
1933	Publication of Tawfīq al-Ḥakīm's play *ahl al-kahf* (*The People of the Cave*).
1938	Publication of Ṭāhā Ḥusayn's book on the future of culture in Egypt (*mustaqbal ath-thaqāfa fī Miṣr*).
1956	Abolition by Egypt of all religious courts and minority parliaments.

Sources

Adham, Ismāʿīl Aḥmad. "bayna al-gharb wa-sh-sharq" [= letter no. 260, June 27, 1938]. In *min maṣādir at-tārīkh al-islāmī wa-nuṣūṣ ukhrā*, Damascus: Dār Bitrā, 2009: 139–45.
Adham, Ismāʿīl Aḥmad. "li-mādhā ana mulḥid?" [1937]. In *min maṣādir al-tārīkh al-islāmī wa-nuṣūṣ ukhrā*, Damascus: Dār Bitrā, 2009: 157–6.
Ainian, Georgios. *Συλλογή ανεκδότων συγγραμμάτων του αιδίμου Ευγενίιου του Βουλγάρεως και τινων άλλων μετατυπωθέντων*. Athens: K. Pallis, 1838.
Amīn, Qāsim. *taḥrīr al-marʾa* [Cairo 1899]. Damascus: Dar al-Baath, s.a.
Aswānī, ʿAlāʾ al-. "hal taṣluḥ ad-dīmuqrāṭīya li-ḥukm al-muslimīn?" In *maqālāt ʿAlāʾ al-Aswānī*, Vol. 2: *hal nastaḥaqq ad-dīmuqrāṭīya?*, 134–7. Cairo: Dar ash-Shuruq [2nd ed.], 2010.
Barqāwī, Aḥmad Nasīm. *muḥāwala fī qirāʾat ʿaṣr an-nahḍa: al-iṣlāḥ ad-dīnī, an-nazʿa al-qawmīya*. Beirut: ar-Ruwwād li-n-nashr wa-t-tawzīʿ, 1988.
Bonaparte, Napoleon. *Napoleons Briefe*, edited by Friedrich Schulze. Leipzig: Insel-Verlag, 1912.
Bustānī, Sulaymān al-. *ʿibra wa-dhikrā aw ad-daula al-ʿuthmānīya qabla al-dastūr wa-baʿdahu*. Beirut: Dar at-Taliʿa, [1904] 1978.
Chevalier, Michel. *Système de la Méditerranée*. Paris: Manucius, 1832.
Diderot, Denis. "Observations sur le Nakaz [1774]." In *Oeuvres politiques*, edited by Paul Vernière. Paris: Garnier Frères, 1963.
Diovouniotis, Konstantinos. *Μητροφάνους Κριτοπούλου ανέκδοτος γραμματική της απλής ελληνικής*. Athens: Makris, 1924.
Dragoumis, Ionos. *Ὁ Ἑλληνισμός μου καί οἱ Ἕλληνες*. Athens: Nea Thesis, [1903–11] 1991.
Gökalp, Ziya. *The Principles of Turkism*. Transl. from the Turkish and annotated by Robert Devereux. Leiden: Brill, 1968.
Hairi, Abdul-Hadi [ʿAbdolhādī Hāʾerī]. "The Idea of Constitutionalism in Persian Literature prior to the 1906 Revolution," in *Akten des VII. Kongresses für Arabistik u. Islamwissenschaft*, edited by Albert Dietrich, 189–207. Göttingen: Vandenhoeck und Ruprecht, 1976.
Ḥakīm, Tawfīq al-. *ahl al-kahf* [1933]. Cairo: Dar ash-Shuruq, 2008.
Hartmann, Martin. *Der Islamische Orient: Berichte und Forschungen*, Vol. III: *Unpolitische Briefe aus der Türkei*. Leipzig: Rudolf Haupt, 1910.
Hatzidakis, George. *Einleitung in die neugriechische Grammatik*. Leipzig 1892, repr. Hildesheim and New York: Olms, 1977.
Herder, Johann Gottfried. *Herders Werke*, edited by Heinrich Meyer, Hans Lambel and Eugen Kühnemann. Berlin and Stuttgart: Union Deutsche Verlagsgesellschaft, 1893.
Hilarion. "Letter to the Patriarch of Constantinople, Anthimos, and the Synod of the Greek Church," publ. in Engl. translation in *The Twenty-First Report of the British and Foreign Bible Society*. London: Augustus Applegath, 1825: 154.
Ḥusayn, Ṭāhā. "niẓām al-Athīniyīn" [1921]. In *al-majmūʿa al-kāmila li-muʾallafāt al-Duktūr Ṭāhā Ḥusayn*, vol. 8, Beirut: Dar al-Kitab al-Lubnani, 1973–1983: 281–384.
Ḥusayn, Ṭāhā. "qādat al-fikr" [Cairo 1925]. In *al-majmūʿa al-kāmila li-muʾallafāt al-Duktūr Ṭāhā Ḥusayn*, vol. 8, Beirut: Dar al-Kitab al-Lubnani 1973–1983: 175–280.
Jowett, William. *Christian Researches in the Mediterranean from 1815 to 1820: In Furtherance of the Objects of the Church Missionary Society*. London: Seeley and Hatchard, 1822.

https://doi.org/10.1515/9783110634006-008

Koraes, Adamantios, de L. R. S. Chardon, and de P. W. Brunet. *Report on the Present State of Civilization in Greece*. In *Lettres inédites de Coray à Chardon de la Rochette*. Paris 1877: 451–90, English translation in *Nationalism in Asia and Africa*, edited by Elie Kedourie, 153–88. Trowbridge and London: Weidenfeld and Nicholson, 1970.

Maurer, Georg Ludwig von. *Das griechische Volk in öffentlicher, kirchlicher und privatrechtlicher Beziehung vor und nach dem Freiheitskampfe bis zum 31. Juli 1834*, 2 vols. Heidelberg: Mohr 1835.

Mūsā, Salāma. *muqaddimat as-subirmān* [1910]. Cairo and Alexandria: Dar al-Mustaqbal, 1977.

Mūsā, Salāma. *mā hiya an-nahḍa?* [1935]. Damascus: Dar al-Baath, 1962.

N.N., "an-Nahḍa al-miṣrīya al-akhīra," in *al-Hilāl*, vol. I, 1/1, Sep. 1, 1892 (Cairo): 123–5.

N.N., "āthār as-Sāmiyīn fī l-madīna l-ūrubīya aw āthār ash-sharq fī madīnat al-ġarb," in *al-Hilāl* no. 6, March 1, 1914: 403–15.

N.N., "al-Yūnān majduhā al-māḍī inbi'āthuhā l-ḥadīth," in *al-Hilāl*, vol. XXIV, No. 4, Jan. 1916: 268–77.

Shahbandar, 'Abdarraḥmān. "at-taṭawwur al-ijtimā'ī wa-s-siyāsī l-ḥadīth fī sh-sharq al-adnā." In *al-maqālāt*, edited by Muḥammad Kāmil al-Khaṭīb, 335–48 [first publ. in al-Muqtaṭaf 2:79 (1931)]. Damascus: Manshurat Wizarat ath-Thaqafa 1993.

Sharafī, Muḥammad [Mohamed Charfi]. *al-Islām wa-l-ḥurrīya: sū' at-tafāhum at-tārīkhī*. Tunis: Dar Petra, 2008.

Shumayyil, Shiblī ash-. *Majmū'a*. Cairo: Muqtataf, 1900–10.

Taqīzādeh, Seyyed Ḥasan. "The History of Modern Iran (Lectures given in Columbia University)." In *Opera Minora: S.H. Taqīzādeh's Articles and Essays*, Vol. VIII: *Unpublished Writings (in European Languages)*, edited by Iraj Afshar, 193–256. Tehran: Shekufan, 1979.

Usṭuwānī, Muḥammad Sa'īd al-. *Mashāhid wa-aḥdāth dimashqīya fī muntaṣaf al-qarn al-tāsi' 'ashar: 1256/1840–1277/1861*, edited by As'ad al-Usṭuwānī. Damascus: Dar al-Jumhuriyya, 1993.

Vélestinlis Ferraios, Rigas. *Documents inédits concernant Rigas Vélestinlis et ses compagnons de Martyre*, edited by Émile Legrand. Paris: E. Leroux, 1892.

Voulgaris, Evgenios. *Γενουηναίου: Στοιχεία της Μεταφυσικής*. Vienna: Georgios Vendotis, 1806.

Zanjānī, Schaykh Ebrāhīm. *khāṭerāt (sar-gozasht-e zendegī-ye man)*, edited by Gholāmḥossein Mīrzā Ṣāleḥ. Tehran: Melli, [2nd ed.] 2001.

Zaqzūq, Muḥammad Ḥamdī. *dawr al-islām fī taṭawwur al-fikr al-falsafī*. Cairo: Maktabat Wahba, 1984.

Zaydān, Jurjī. *riḥla ilā Ūrubbā sanat 1912* [1912]. Cairo: al-Hilal, 1923.

Literature

Abdel Meguid, Ibrahim. *Birds of Amber.* Cairo: American University in Cairo Press, 2005.

Abū Zayd, Naṣr Ḥāmid. *al-Imām aš-Šāfiʿī wa-taʾsīs al-īdīyulūǧīya al-wasaṭīya.* Cairo: Sina li-n-nashr, 1992.

Acemoglu, Daron, and James A. Robinson. *Why Nations Fail: The Origins of Power, Prosperity, and Poverty.* New York: Crown Business, 2012.

Adorno, Theodor W., and Max Horkheimer. *Dialektik der Aufklärung.* Frankfurt/Main: Suhrkamp, [1947] 2003.

Ajami, Fouad. *The Dream Palace of the Arabs: A Generation's Odyssey.* New York: Vintage Books, 1999.

Altaner, Berthold. "Zur Geschichte der anti-islamischen Polemik während des 13. und 14. Jahrhunderts." *Historisches Jahrbuch* 56 (1936): 229–30.

Anderson, Benedict. *Die Erfindung der Nation. Zur Karriere eines folgenreichen Konzepts.* Frankfurt/Main and New York: Campus, 1993.

Anderson, J.N.D. "Law Reform in Egypt, 1850–1950." In *Revolution in the Middle East and Other Case Studies*, edited by P.J. Vatikiotis, 146–172. London: Allen and Unwin, 1972.

Arnzen, Rüdiger. *Platonische Ideen in der arabischen Philosophie: Texte und Materialien zur Begriffsgeschichte von ṣuwar aflāṭūniyya und muthul aflāṭūniyya.* Berlin: De Gruyter, 2011.

Arsuzi-Elamir, Dalal. *Arabischer Nationalismus in Syrien: Zaki al-Arsuzi und die arabisch-nationale Bewegung an der Peripherie Alexandretta/Antakya 1930–1938.* Münster et al.: Lit, 2003.

Badran, Abū al-Fadl Muḥammad. "'...denn die Vernunft ist ein Prophet' – Zweifel bei Abū 'l-ʿAlāʾ al-Maʿarrī." In *Atheismus im Mittelalter und in der Renaissance*, edited by Friedrich Niewöhner and Olaf Pluta, 61–84. Wiesbaden: Harrassowitz, 1999.

Bagster, Samuel. *The Bible of Every Land: A History of the Sacred Scriptures in Every Language and Dialect Into Which Translations Have Been Made.* London: *The Bible of Every Land*, 1848.

Bauer, Thomas. *Die Kultur der Ambiguität: Eine andere Geschichte des Islams.* Berlin: Verlag der Weltreligionen, 2011.

Bayat, Mangol. *Iran's First Revolution: Shiism and the Constitutional Revolution of 1905–1909.* Oxford: Oxford University Press, 1991.

Beck, Hans-Georg. "Humanismus und Palamismus," in *Actes du XIIe congrès internatioonal d'études byzantines, Ochride, 10–16 septembre 1961*, tome 1, Belgrade 1963 : 63–82.

Beck, Hans-Georg. *Das byzantinische Jahrtausend.* Munich: C.H. Beck, 1994.

Berlin, Isaiah. "Two Concepts of Liberty." In *Four Essays on Liberty*, 118–72. (London, Oxford and New York: Oxford University Press, 1969.

Berlin, Isaiah, and Ramin Jahanbegloo. *Den Ideen die Stimme zurückgeben: Eine intellektuelle Biographie in Gesprächen.* Frankfurt/Main: Suhrkamp, 1994.

Binswanger, Karl. "Türkei." In *Der Islam in der Gegenwart: Entwicklung und Ausbreitung. Staat, Politik und Recht, Kultur und Religion*, edited by Werner Ende and Udo Steinbach, 212–20. Munich: C.H. Beck, 1991.

Bleek, Wilhelm. *Geschichte der Politikwissenschaft in Deutschland.* Munich: C.H. Beck, 2001.

Blum, Paul Richard. *Philosophieren in der Renaissance.* Stuttgart: Kohlhammer, 2004.

https://doi.org/10.1515/9783110634006-009

Blumenberg, Hans. *Das Lachen der Thrakerin: Eine Urgeschichte der Theorie.* Frankfurt/Main: Suhrkamp, 1987.
Blumenberg, Hans. *Arbeit am Mythos.* Frankfurt/Main: Suhrkamp, 1996.
Blumenberg, Hans. *Die Legitimität der Neuzeit.* Frankfurt/Main: Suhrkamp, 1996.
Blumenberg, Hans. *Höhlenausgänge.* Frankfurt/Main: Suhrkamp, 1996.
Blumenberg, Hans. *Wirklichkeiten in denen wir leben: Aufsätze und eine Rede.* Stuttgart: Reclam, 1999.
Blumenberg, Hans. *Die Vollzähligkeit der Sterne.* Frankfurt/Main: Suhrkamp, 2000.
Blumenberg, Hans. "Wirklichkeitsbegriff und Wirkungspotential des Mythos." In *Terror und Spiel*, edited by Manfred Fuhrmann, Munich: W. Fink 1971: 11–66, repr. in *Ästhetische und metaphorologische Schriften*, edited by Anselm Haverkamp, 327–405. Frankfurt/Main: Suhrkamp, 2001.
Blumenberg, Hans. *Lebenszeit und Weltzeit.* Frankfurt/Main: Suhrkamp, 2001.
Blumenberg, Hans. *Präfiguration: Arbeit am politischen Mythos*, edited by Angus Nicholls and Felix Heidenreich. Berlin: Suhrkamp, 2014.
Bobzin, Hartmut. *Der Koran im Zeitalter der Reformation.* Beirut and Wiesbaden: Harrassowitz, 2008.
Brockelmann, Carl. *Geschichte der islamischen Völker und Staaten.* Munich and Berlin, 1943, repr. Hildesheim and New York: Georg Olms, 1977.
Brugman, Jan, and Frank Schröder. *Arabic Studies in the Netherlands.* Leiden: Brill, 1979.
Bubner, Rüdiger. *Welche Rationalität bekommt der Gesellschaft? Vier Kapitel aus dem Naturrecht.* Frankfurt/Main: Suhrkamp, 1996.
Buchenau, Klaus. *Auf russischen Spuren: Orthodoxe Antiwestler in Serbien, 1850–1945.* Wiesbaden: Harrassowitz, 2011.
Burke, Peter. "Did Europe exist before 1700?," in *History of European Ideas*, Vol. 1/1 (1980): 21–29.
Cachia, Pierre. *Ṭāḥā Ḥusayn: His Place in the Egyptian Literary Renaissance.* London: Luzac, 1956.
Campenhausen, Hans. *Griechische Kirchenväter.* Stuttgart, Berlin and Cologne: Kohlhammer, 1993.
Cassirer, Ernst. *Die Idee der republikanischen Verfassung: Rede zur Verfassungsfeier am 11. August 1928.* Hamburg: De Gruyter, 1929.
Cassirer, Ernst. *The Myth of the State.* New Haven: Yale University Press, 1946.
Cassirer, Ernst. *Nachgelassene Manuskripte und Texte*, edited by Klaus Christian Köhnke, John Michael Krois and Oswald Schwemmer, Vol. 3: *Geschichte, Mythos. Mit Beilagen: Biologie, Ethik, Form, Kategorienlehre, Kunst, Organologie, Sinn, Sprache, Zeit*, edited by Klaus Christian Köhnke, Herbert Kopp-Oberstebrink and Rüdiger Kramme. Hamburg: Felix Meiner, 2002.
Chaconas, Stephen. "The Jefferson–Korais Correspondence," in *The Journal of Modern History*, Vol. 14, No. 1 (March 1942): 64–70.
Clark, Christopher. *Sleepwalkers: How Europe Went to War in 1914.* London: Penguin, 2013.
Clark, Gregory. *The Son Also Rises: Surnames and the History of Social Mobility.* Princeton und Oxford: Princeton University Press, 2014.
Clauss, Manfred. "Die συμφών von Kirche und Staat zur Zeit Justinians." In *Klassisches Altertum, Spätantike und frühes Christentum: Adolf Lippold zum 65. Geburtstag*

gewidmet, edited by Karlheinz Dietz, Dieter Hennig, and Hans Kaletsch, 579–93. Würzburg: Der christliche Osten, 1993.

Clayer, Nathalie. "Der Balkan, Europa und der Islam." In *Wieser Enzyklopädie des Europäischen Ostens, Vold. 11: Europa und die Grenzen im Kopf*, edited by Karl Kaser, Dagmar Gramshammer-Kohl, and Robert Pichler, 303–30. Klagenfurt, Vienna and Ljubljana: Wieser, 2003.

Clogg, Richard. "The ‚Dhidhaskalia Patriki' (1798): An Orthodox Reaction to French Revolutionary Propaganda," in *Middle Eastern Studies*, Vol. 5, No. 2 (May, 1969): 87–115.

Cohen, Floris. *Die zweite Erschaffung der Welt*. Frankfurt/Main and New York: Campus, 2010.

Cole, Juan. *Napoleon's Egypt: Invading the Middle East*. New York et al.: Palgrave Macmillan, 2008.

Colley, Linda. *Britons: Forging the Nation 1707–1837*. New Haven and London: Yale University Press, 2009.

Cook, Michael. *Ancient Religions, Modern Politics: The Islamic Case in Comparative Perspective*. Princeton/New Jersey: Princeton University Press, 2014.

Darwin, John. *Der Imperiale Traum: Die Globalgeschichte großer Reiche 1400–2000*. Frankfurt/Main and New York: Campus, 2010.

Dawson, Christopher. *Die Gestaltung des Abendlandes: Eine Einführung in die Geschichte der abendländischen Einheit*. Leipzig: Jakob Hegner, 1935.

Demos, Raphael. "The Neo-Hellenic Enlightenment (1750–1821)," in *Journal of the History of Ideas*, Vol. 19, No. 4 (1958): 523–41.

Dimaras, Konstantinos. *Νεοελληνικός Διαφωτισμός*. Athens: Ermis, 2002.

Dīk, Aghnāṭiyūs. *al-masīḥīya fī Sūrīa: tārīkh wa-ish'ā'*. 3 vols., Aleppo: Dar al-Kitab al-Muqaddas, 2008–9.

Diner, Dan. "Zweierlei Osten: Europa zwischen Westen, Byzanz und Islam." In *Das Europa der Religionen*, edited by Otto Kallscheuer, 97–113. Frankfurt/Main: S. Fischer, 1996.

Dölger, Franz. *Der griechische Barlaam-Roman: Ein Werk des H. Johannes von Damaskos*. Ettal: Buch-Kunstverlag, 1953.

Duchhardt, Heinz. *Der Wiener Kongress: Die Neugestaltung Europas 1814/15*. Munich: C.H. Beck, 2013.

Ebert, Johannes. *Religion und Reform in der arabischen Provinz: Ḥusayn Al-Ǧisr aṭ-Ṭarābulusī (1845–1909): Ein Islamischer Gelehrter zwischen Tradition und Reform*. Frankfurt/Main: Peter Lang, 1991.

Elger, Ralf. "Selbstdarstellungen aus Bilâd ash-Shâm. Überlegungen zur Innovation in der arabischen autobiographischen Literatur im 16. und 17. Jahrhundert." In *Eigene und fremde Frühe Neuzeiten: Genese und Geltung eines Epochenbegriffs*, edited by Renate Dürr, Gisela Engel, and Johannes Süßmann [= *Historische Zeitschrift*, Beiheft 35], 123–37. Munich: R. Oldenbourg, 2003.

Endreß, Gerhard. "Der Islam und die Einheit des mediterranean Kulturraums im Mittelalter." In *Das Mittelmeer: Die Wiege der europäischen Kultur*, edited by Klaus Rosen, 270–95. Bonn: Bouvier, 1998.

Fähndrich, Hartmut. "Orientalismus und Orientalismus: Überlegungen zu Edward Said, Michel Foucault und westlichen ‚Islamstudien.'" In *Gegenwart als Geschichte: Islamwissenschaftliche Studien: Fritz Steppat zum fünfundsechzigsten Geburtstag*, edited by Axel Havemann and Baber Johansen, 178–86. Leiden et al.: Brill, 1988.

Feroz, Ahmad. *The Making of Modern Turkey.* London and New York: Routledge, 1994.
Flasch, Kurt. *Aufklärung im Mittelalter? Die Verurteilung von 1277: Das Dokument des Bischofs von Paris übersetzt und erklärt.* Mainz: Dieterich, 1989.
Frazee, Charles. *Catholics and Sultans: The Church and the Ottoman Empire 1453–1923.* Cambridge et al.: Cambridge University Press, 2006.
Freitag, Ulrike. "The Critique of Orientalism." In *Companion to Historiography*, edited by M. Bentley, 620–38. London and New York: Routledge, 1997,.
Fück, Johann. *Die arabischen Studien in Europa bis in den Anfang des 20. Jahrhunderts.* Leipzig: Harrassowitz, 1995.
Francis Fukuyama. *The End of History and the Last Man.* London: Hamish Hamilton, 1992.
Francis Fukuyama. *Political Order and Political Decay: From the French Revolution to the Present.* London: Profile Books, 2014.
Gauß, Julia. "Glaubensdiskussionen zwischen Ostkirche und Islam im 8.–11. Jahrhundert," in *Theologische Zeitschrift* 19 (1963): 14–28.
Gershoni, Israel, and James P. Jankowski. *Egypt, Islam, and the Arabs: The Search for Egyptian Nationhood, 1900–1930.* New York and Oxford: Oxford University Press, 1986.
Goldziher, Ignaz. *Die Richtungen der islamischen Koranauslegung.* Leiden: Brill, 1970.
Gollwitzer, Heinz. *Europabild und Europagedanke: Beiträge zur deutschen Geistesgeschichte des 18. und 19. Jahrhunderts.* Munich: C.H. Beck, 1964.
Gounaris, Basil. "Social cleavages and national ‚awakening' in Ottoman Macedonia," in *East European Quarterly* XXIX, No. 4 (1996): 409–2.
Grunebaum, Gustave. "The Problem of Cultural Influence", in *Charisteria Orientalia praecipue as Persiam pertinentia*, edited by Felix Tauer, Věra Kubíčková, and Ivan Hrbek, 86–99. Prague: Ceskoslovenska Akademie, 1956.
Guariglia, Osvaldo. *Universalismus und Neuaristotelismus in der zeitgenössischen Ethik.* Hildesheim et al.: Georg Olms, 1995.
Gumbrecht, Hans Ulrich. *Diesseits der Hermeneutik: Die Produktion von Präsenz.* Frankfurt/Main: Suhrkamp, 2004.
Günter, Andrea. *Weibliche Autorität, Freiheit und Geschlechterdifferenz: Bausteine einer feministischen politischen Theorie.* Königstein/ Taunus: Ulrike Helmer Verlag, 1996.
Haarmann, Ulrich. "Ideology and history, identity and alterity: The Arab image of the Turk from the 'Abbasids to modern Egypt," in *International Journal of Middle East Studies*, Vol. 20 (1988): 175–96.
Haffner, Sebastian. "Das Nachleben Roms," in *Historische Variationen*, 31–8. Stuttgart and Munich: DVA, 2001.
Hagen, Gottfried, and Tilman Seidensticker. "Reinhard Schulzes Hypothese einer islamischen Aufklärung. Kritik einer historiographischen Kritik," in *Zeitschrift der Deutschen Morgenländischen Gesellschaft*, Vol. 148 (1998): 38–110.
Hanf, Theodor. "Die christlichen Gemeinschaften im gesellschaftlichen Wandel des arabischen Vorderen Orients," in *Orient* 1 /1981; 29–49.
Harel, Y. "Midhat Pasha and the Jewish Community of Damascus: Two New Documents," in *Turcica* 28 (1996): 339–46.
Harlan, David. "Der Stand der Geistesgeschichte und die Wiederkehr der Literatur." In *Die Cambridge School der politischen Ideengeschichte*, edited by Martin Mulsow and Andreas Mahler, 155–202. Berlin: Suhrkamp, 2010.

Hassiotis, I. K. "From the ‚Refledging' to the ‚Illumination of the Nation:' Aspects of Political Ideology in the Greek Church under Ottoman Domination," in *Balkan Studies*, Vol. 40, No. 1 (1999): 41–55.

Havemann, Axel. "Geschichte und Geschichtsschreibung im Libanon: Kamāl Ṣalībī und die nationale Identität." In *Gegenwart als Geschichte: Islamwissenschaftliche Studien. Fritz Steppat zum fünfundsechzigsten Geburtstag*, edited by Axel Havemann and Baber Johansen, 225–43. Leiden et al.: Brill, 1988.

Hefny, Assem. *Herrschaft und Islam: Religiös-politische Termini im Verständnis ägyptischer Autoren.* Frankfurt/ Main: Peter Lang, 2014.

Henrich, Günther. "Als Denker Archaist, als Dichter auch Demotizist: Zu Vúlgaris' Paraphrase des Voltaireschen Memnon." In *Evgenios Vulgaris und die neugriechische Aufklärung in Leipzig: Konferenz an der Universität Leipzig, 16.–-18. Oktober 1996*, edited by Günther S. Henrich, 99–111. Leipzig: Leipziger Universitätsverlag, 2003.

Herf, Jeffrey. *Nazi Propaganda for the Arab World.* New Haven/ Conn.: Yale University Press, 2009.

Hering, Gunnar. *Ökumenisches Patriarchat und europäische Politik, 1620–1638.* Wiesbaden: Steiner, 1968.

Hering, Gunnar. "Die Auseinandersetzungen über die neugriechische Schriftsprache." In *Sprachen und Nationen im Balkanraum: Die historischen Bedingungen der Entstehung der heutigen Nationalsprachen*, edited by Christian Hannick, 125–94. Cologne: Böhlau, 1987.

Herman, Arthur. *How the Scots Invented the Modern World: The True Story of How Western Europe's Poorest Nation Created Our World & Everything in it.* New York: Crown, 2001.

Herzog, Christoph. "Zum Niedergangsdiskurs im Osmanischen Reich." In *Mythen, Geschichte(n), Identitäten: Der Kampf um die Vergangenheit*, edited by Stephan Conermann, 69–90. Hamburg: EB-Verlag 1999.

Herzog, Christoph. "Aufklärung und Osmanisches Reich: Annäherung an ein historiographisches Problem." In *Die Aufklärung und ihre Weltwirkung*, edited by Wolfgang Hardtwig, 291–321. Göttingen: Vandenhoeck Ruprecht, 2010.

Heyer, Andreas. *Materialien zum politischen Denken Diderots: Eine Werksmonographie.* Hamburg: Kovač, 2004.

Heyer, Friedrich. *Die Orientalische Frage im kirchlichen Lebenskreis: Das Einwirken der Kirchen des Auslands auf die Emanzipation der orthodoxen Nationen Südosteuropas 1804–1912.* Wiesbaden: Harrassowitz, 1991.

Heywood, Colin. "The Frontier in Ottoman History: Old Ideas and New Myths." In *Writing Ottoman History: Documents and Interpretations.* Aldershot: Ashgate 2002: I, 238–9.

Heyworth-Dunne, James. *An Introduction to the History of Education in Modern Egypt.* London: F. Cass, 1968.

Himmelfarb, Getrude. *The Roads to Modernity: The British, French, and American Enlightenments.* New York: Vintage, 2004.

Höfert, Almut. *Den Feind beschreiben: ‚Türkengefahr' und europäisches Wissen über das Osmanische Reich, 1450–1600.* Frankfurt/Main and New York: Campus, 2003.

Hösch, Edgar. *Geschichte der Balkanländer: Von der Frühzeit bis zur Gegenwart.* Munich: C.H. Beck, 2002.

Hosking, Geoffrey. *Russland: Nation und Imperium, 1552–1917.* Berlin: Berliner Taschenbuch Verlag, 2003.

Hourani, Albert. *Arabic Thought in the Liberal Age: 1798–1939.* Cambridge: Cambridge University Press, 2002.
Hourani, Albert. "Kultur und Wandel: Der Nahe und Mittlere Osten im 18. Jahrhundert." In *Der Islam im europäischen Denken.* Frankfurt/Main: S. Fischer, 1994: 169–201.
Hourani, Albert. "Islamische Geschichte, Geschichte des Nahen und Mittleren Ostens, moderne Geschichte." In *Der Islam im europäischen Denken.* Frankfurt/Main: S. Fischer, 1994: 109–41.
Huntington, Samuel. *The Clash of Civilizations and the Remaking of World Order.* New York: Foreign Affairs, 1996.
Akira Iriye, "Japan's drive to great-power status." In *The Emergence of Meiji Japan*, edited Marius B. Jansen, 268–329. Cambridge: Cambridge University Press, 1995.
Irmscher, Johannes. "Über den Morgenland-Begriff." In *Byzantino-Sicula II. Miscellanea di scritti in memoria di G. Rossi Taibbi*, 295–300. Palermo: Istituto siciliano di studi bizantini e neoellenici, 1975.
Israel, Jonathan. *Enlightenment Contested: Philosophy, Modernity, and the Emancipation of Man: 1670–1752.* Oxford: Oxford University Press, 2006.
Jenkins, Romilly. *Byzantium: The Imperial Centuries, AD 610–1071.* Toronto: University of Toronto Press, 1987.
Kant, Immanuel. "Beantwortung der Frage: Was ist Aufklärung" [1784]. In *Die Kritiken*, 633–40. Frankfurt/Main: Zweitausendeins, 2008.
Karpat, Kemal. *The Politicization of Islam: Reconstructing Identity, State, Faith, and Community in the Late Ottoman State.* Oxford: Oxford University Press, 2001.
Karpat, Kemal. "Millets and Nationality: The Roots of the Incongruity of Nation an State in the Post-Ottoman Era." In *Christians and Jews in the Ottoman Empire*, edited by Benjamin Braude and Bernard Lewis, New York, 1982, repr. in Kemal Karpat, *Studies on Ottoman Social and Political History: Selected Articles and Essays*, Leiden et al., 2002: 611–46.
Karsh, Efraim. *Islamic Imperialism: A History.* New Haven and London: Yale University Press, 2006.
Karsh, Efraim, and Inari Karsh. *Empires of the Sand: The Struggle for Mastery in the Middle East, 1789–1923.* Cambridge/Mass. and London: Harvard University Press, 2001.
Kaser, Karl. *The Balkans and the Near East: Introduction to a Shared History.* Vienna and Berlin: Lit, 2011.
Kaufman, Asher. "Phoenicianism: The Formation of an Identity in Lebanon in 1920," in *Middle Eastern Studies*, Vol. 37, No. 1, January 2001: 173–94.
Kaufman, Asher. *Reviving Phoenicia: In Search for Identity in Lebanon.* London: Tauris, 2004.
Kaufmann, Thomas. "Kontinuitäten und Transformationen im okzidentalen Islambild des 15. und 16. Jahrhunderts." In *Judaism, Christianity, and Islam in the Course of History: Exchange and Conflicts*, edited by Lothar Gall and Dietmar Willoweit, 287–306. Munich: R. Oldenbourg 2011.
Kaufmann, Thomas. *Der Anfang der Reformation: Studien zur Kontextualität der Theologie, Publizistik und Inszenierung Luthers und der reformatorischen Bewegung.* Tübingen: Mohr Siebeck, 2012.
Kaufmann, Thomas. "Kontinuitäten und Transformationen im okzidentalen Islambild des 15. und 16. Jahrhunderts." In *Judaism, Christianity, and Islam in the Course of History:*

Exchange and Conflicts, edited by Lothar Gall and Dietmar Willoweit, 287–306. Munich: R. Oldenbourg, 2011.

Kelleter, Frank. *Amerikanische Aufklärung: Sprachen der Rationalität im Zeitalter der Revolution*. Paderborn: Schöningh, 2002.

Khawaja, Irfan. "Essentialism, Consistency and Islam: A Critique of Edward Said's Orientalism." In *Postcolonial Theory and the Arab-Israel Conflict*, edited by Philip Carl Salzman and Donna Robinson Divine, 12–36. London and New York: Routledge, 2008.

Kitromilides, Paschalis M. *Enlightenment and Revolution: The Making of Modern Greece*. Cambridge/Mass. and London: Harvard University Press, 2013.

Klein, Dietrich. "Hugo Grotius' position on Islam as described in De Veritate Religionis Christianae, Liber VI." In *Socinianism and Arminianism: Antitrinitarians, Calvinists and Cultural Exchange in Seventeenth-Century Europe*, edited by Martin Mulsow and Jan Rohls, 149–73. Leiden und Boston: Brill, 2005.

Knapp, Martin. *Evjenios Vulgaris im Einfluss der Aufklärung: der Begriff der Toleranz bei Vulgaris und Voltaire*. Amsterdam: Hakkert, 1984.

Kohl, Karl-Heinz. *Abwehr und Verlangen: Zur Geschichte der Ethnologie*. Frankfurt/Main and New York: Campus, 1987.

Kohn, Hans. *A History of Nationalism in the East*. New York: Harcourt, Brace and Co., 1929.

Koselleck, Reinhart. *Kritik und Krise: Ein Beitrag zur Pathogenese der bürgerlichen Welt*. Freiburg and Munich: K. Alber, 1959.

Kozlarek, Oliver. *Moderne als Weltbewusstsein: Ideen für eine humanistische Sozialtheorie in der globalen Moderne*. Bielefeld: transcript, 2011.

Kreutz, Michael. *Das klassische griechische Schrifttum in der Rezeption der arabischen Nahḍa*. Unpublished M.A. thesis, Ruhr-University of Bochum. Bochum 1999.

Kreutz, Michael. "Sulaymān al-Bustānīs Arabische Ilias. Ein Beispiel für arabischen Philhellenismus im ausgehenden Osmanischen Reich," in *Die Welt des Islams* 44, 2 (2004): 155–94.

Kreutz, Michael. "The Greek Classics in Modern Middle Eastern Thought." In *Judaism, Christianity, and Islam in the Course of History: Exchange and Conflicts*, edited by Lothar Gall and Dietmar Willoweit, 77–92. Munich: R. Oldenbourg, 2011.

Kreutz, Michael. "Unrest at the Gates of Aleppo: British Perspectives on Bedouins in Northern Syria, 1848–1913," in *Journal of Levantine Studies* Vol. 2/2 (2012): 105–29.

Kreutz, Michael. "Empire and Enlightenment: Greek Poetry in Ottoman Letters." In *Marginal Perspectives on Early Modern Ottoman Culture: Missionaries, Travellers, Booksellers*, edited by Ralf Elger and Ute Pietruschka [= Hallesche Beiträge zur Orientwissenschaft, Vol. 23/ 2013], 1–15. Halle (Saale): ZRS, 2013.

Kreutz, Michael. *Das Ende des levantinischen Zeitalters: Europa und die Östliche Mittelmeerwelt, 1821–1939*. Hamburg: Kovac, 2013.

Kreutz, Michael. *Zwischen Religion und Politik: Die verschlungenen Pfade der Moderne*. Bochum: Verlag Michael Kreutz, 2016.

Kreutz, Michael. "Ideengeschichte als Blindstelle historischer Forschung in Deutschland," in *michaelkreutz.net*, Aug. 16, 2017, URL= http://www.michaelkreutz.net/2017/ideengeschichte-als-leerstelle-der-heutigen-forschung/

Kuntz, Marion. *Guillaume Postel: Prophet of the Restitution of All Things: His Life and Thought*. The Hague: M. Nijhoff, 1981.

Kuzmics, Helmut, and Roland Axtmann. *Autorität, Staat und Nationalcharakter: Der Zivilisationsprozeß in Österreich und England, 1700–1900.* Opladen: Leske und Budrich, 2000.

Landau, Jacob M. "Prolegomena to a Study of Secret Societies in Modern Egypt." In *Middle Eastern Themes: Papers in History and Politics.* London 1973 [= reprint from *Middle Eastern Studies*, I (2), Jan. 1965, 1–52): 7–56.

Lapidus, Ira M. "Islamisches Sektierertum und das Rekonstruktions. und Umgestaltungspotential der islamischen Kultur." In *Kulturen der Achsenzeit*, Bd. II: *Ihre institutionelle und kulturelle Dynamik*, Teil 3: *Buddhismus, Islam, Altägypten, westliche Kultur*, edited by Shmuel N. Eisenstadt, 161–88. Frankfurt/Main 1992.

Lazarus-Yafeh, Hava. "Die islamische Reaktion auf den Rationalismus." In *Kulturen der Achsenzeit*, Bd. II: *Ihre institutionelle und kulturelle Dynamik*, Teil 3: *Buddhismus, Islam, Altägypten, westliche Kultur*, edited by Shmuel N. Eisenstadt, 210–25. Frankfurt/Main 1992.

Lewis, Bernard. "The Idea of Political Freedom in Modern Islamic Political Thought." In *Islam in History: Ideas, Men and Events in the Middle East*, London: Alcove Press 1973: 267–281.

Lewis, Bernard. *The Multiple Identitites of the Middle East.* New York: Schocken Books, 1998.

Lewis, Bernard. "On Occidentalism and Orientalism," in idem, *From Babel to Dragomans: Interpreting the Middle East.* London: Weidenfeld & Nicolson 2004: 430–3.

Lewis, Bernard. "From Pilgrims to Tourists: A Survey of Middle Eastern Travel," in idem, *From Babel to Dragomans: Interpreting the Middle East.* London: Weidenfeld & Nicolson, 2004; 137–51.

Lewis, Bernard. *Faith and Power: Religion and Politics in the Middle East.* Oxford: Oxford University Press, 2010.

Lewis, Bernard, and Buntzie E. Churchill. *Notes on a Century: Reflections of a Middle East Historian.* New York: Viking, 2012.

Lill, Rudolf. *Geschichte Italiens vom 16. Jahrhundert bis zu den Anfängen des Faschismus.* Darmstadt: Wissenschaftliche Buchgesellschaft, 1980.

Mackridge, Peter. *Language and National Identity in Greece, 1766–1976.* Oxford: Oxford University Press, 2010.

Maher, Mustafa. "Umrisse einer neuen Kulturphilosophie in Ägypten seit dem 19. Jahrhundert." In *Gegenwart als Geschichte: Islamwissenschaftliche Studien. Fritz Steppat zum fünfundsechzigsten Geburtstag*, edited by Axel Havemann and Baber Johansen, 309–18. Leiden et al.: Brill, 1988.

Maier, Franz Georg. *Die Verwandlung der Mittelmeerwelt* [= Fischer Weltgeschichte Vol. 9]. Frankfurt/ Main: S. Fischer, 1968.

Makrides, Vasilios. "Orthodoxes Ost- und Südosteuropa: Ausnahmefall oder Besonderheit?" In *Die Vielfalt Europas: Identitäten und Räume: Beiträge einer internationalen Konferenz*, Leipzig, 6. bis 9. Juni 2007, edited by Winfried Eberhard, 203–18. Leipzig: Leipziger Universitäts-Verlag, 2009.

Makrides, Vasilios. "Orthodox Anti-Westernism Today: A Hindrance to European Integration?" in *International Journal for the Study of the Christian Church*, vol. 9 (3) (2009): 209–224.

Marx, Michael. "Europa, Islam und Koran: Zu einigen Elementen in der gegenwärtigen gesellschaftlichen Debatte." In *Gehört der Islam zu Deutschland? Fakten und Analysen*

zu einem Meinungsstreit, edited by Klaus Spenlen, 61–98. Düsseldorf: Düsseldorf University Press, 2013.

Massie, Robert K. *Catherine the Great: Portrait of a Woman*. New York: Random House, 2011.

Matl, Josef. *Südslawische Studien*. Munich: R. Oldenbourg, 1965.

Matuz, Josef. *Das Osmanische Reich: Grundlinien seiner Geschichte*. Darmstadt: Wissenschaftliche Buchgesellschaft, 1994.

Mazur, Peter A. "A Mediterranean Port in the Confessional Age: Religious Minorities in Early Modern Naples." In *A Companion to Early Modern Naples*, edited by Tommaso Astarita, 235–56. Leiden: Brill, 2013.

Merkel, Rudolf Franz. "Der Islam im Wandel abendländischen Verstehens," in *Studi e Materiali di Storia delle Religioni* 13 (1937): 68–101.

Mishra, Pankaj. *From the Ruins of Empire: The Intellectuals Who Remade Asia*. New York: Picador, 2012.

Mitsiou, Ekaterini. "Interaktion zwischen Kaiser und Patriarch im Spiegel des Patriarchatsregisters von Konstantinopel." In *Zwei Sonnen am Goldenen Horn? Kaiserliche und patriarchale Macht im byzantinischen Mittelalter: Akten der internationalen Tagung vom 3. bis 5. November 2010*, edited by Michael Grünbart, Lutz Rickelt, and Martin Marko Vučetić, 79–96. Berlin: Lit, 2011.

Moore, James. "The two systems of Francis Hutcheson: On the Origins of Scottish Enlightenment." In *Studies in the Philosophy of Scottish Enlightenment*, edited by Michael Stewart, 37–60. Oxford: Clarendon Press, 2000.

Mujais, Salim [Salīm Maǧāʿis]. *Antoun Saadeh: A Biography*, vol. 1: *The Youth Years*. Beirut: Kutub, 2004.

Mühlpfordt, Günter. "Hellas als Wegweiser zur Demokratie. Griechenmodell und Griechenkritik radikaler Aufklärer–Antikerezeption im Dienst bürgerlicher Umgestaltung." In *Griechenland–Byzanz–Europa: Ein Studienband*, edited by Joachim Herrmann, Helga Köpstein and Reimar Müller, 225–69. Amsterdam: Gieben, 1988.

Müller, Gerhard, ed. *Theologische Realenzyklopädie*, Berlin: De Gruyter, 1993.

Nagel, Tilman. "Gab es in der islamischen Geschichte Ansätze zu einer Säkularisierung?" In *Studien zur Geschichte und Kultur des Vorderen Orients: Festschrift für Bertold Spuler zum siebzigsten Geburtstag*, edited by Hans R. Roemer and Albrecht Noth, 275–88. Leiden: Brill, 1981.

Nagel, Tilman. *Die Festung des Glaubens: Triumph und Scheitern des islamischen Rationalismus im 11. Jahrhundert*. Munich: C.H. Beck, 1988.

Nagel, Tilman. *Geschichte der islamischen Theologie: Von Mohammed bis zur Gegenwart*. Munich: C.H. Beck, 1994.

Nagel, Tilman. *Zu den Grundlagen des islamischen Rechts*. Baden-Baden: Nomos, 2012.

Nagel, Tilman. *Angst vor Allah? Auseinandersetzungen mit dem Islam*. Berlin: Duncker und Humblot, 2014.

Nestle, Wilhelm. *Griechische Weltanschauung in ihrer Bedeutung für die Gegenwart: Vorträge und Abhandlungen*. Stuttgart: Hannsmann, 1946.

Nietzsche, Friedrich. "Zur Genealogie der Moral." In *Sämtliche Werke: kritische Studienausgabe in 15 Bänden*, Vol. 5: *Jenseits von Gut und Böse; Zur Genealogie der Moral*, edited by Giorgio Colli and Mazzino Montinari, 245–412. Munich: dtv and De Gruyter, [1988/1999] 2012.

Nolte, Paul. *Transatlantische Ambivalenzen: Studien zur Sozial- und Ideengeschichte des 18. bis 20. Jahrhunderts.* Berlin: De Gruyter, 2014.

Noutsos, Panagiotis. "Christian Wolff und die neugriechische Aufklärung." In *Evgenios Vulgaris und die neugriechische Aufklärung in Leipzig: Konferenz an der Universität Leipzig, 16.–18. Oktober 19965*, edited by Günther S. Henrich, 76–82. Leipzig: Leipziger Universitätsverlag, 2003.

O'Donnell, Kathleen Ann. "How Twentieth Century Greek Scholars Influenced the Works of Nineteenth Century Greek Translators of 'The Poems of Ossian' by James Macpherson," in *Athens Journal of Philology* Vol. 1, No. 4 (December 2014): 273–84.

O'Donnell, Kathleen Ann. "The Disintegration of the Democratic Eastern Federation and the Demise of its supporters 1885–1896 and the Poems of Ossian," in *ATINER'S Conference Paper Series.* Athens 2015: 3–14.

Odysseus [= Charles Eliot]. *Turkey in Europe.* London: Edward Arnold, 1900.

Oehler, Klaus. *Blicke aus dem Philosophenturm: Eine Rückschau.* Hildesheim, Zürich, New York: Georg Olms, 2007.

Osterhammel, Jürgen. "Welten des Kolonialismus im Zeitalter der Aufklärung." In *Das Europa der Aufklärung und die aussereuropäische koloniale Welt*, edited by Hans-Jürgen Lüsebrink, 19–36. Göttingen: Wallstein, 2006.

Osterhammel, Jürgen. *Die Entzauberung Asiens: Europa und die Asiatischen Reiche im 18. Jahrhundert.* Munich: C.H. Beck, 2010.

Oz-Salzberger, Fania. "Freiheit und die ‚Gemeinschaft aller' in der schottischen Aufklärung." In *Kollektive Freiheitsvorstellungen im frühneuzeitlichen Europa (1400–1850)*, edited by Georg Schmidt, Martin van Gelderen, and Christopher Snigula, 419–29. Frankfurt/Main et al.: Peter Lang 2006.

Papoulia, Basiliki. *Από την αυτοκρατορία στο εθνηκό κράτος.* Thess. and Athens: Ekdoseis Banias, 2003.

Petropoulou, Evi. *Geschichte der neugriechischen Literatur.* Frankfurt/Main: Suhrkamp, 2001.

Petropoulou, Ioanna. "From West to East: The Translation Bridge. An Approach from a Western Perspective." In *Ways to Modernity in Greece and Turkey*, edited by Anna Frangoudaki and Çağlar Keyder, 91–112. London: I.B. Tauris, 2007.

Petsios, Kostas. "Kants Kategorienlehre im Werk von Athanasios Psalidas." In *Evgenios Vulgaris und die neugriechische Aufklärung in Leipzig: Konferenz an der Universität Leipzig, 16.–18. Oktober 1996*, edited by Günther S. Henrich, 55–67 Leipzig: Leipziger Universitätsverlag, 2003.

Peyfuss, Max Demeter. "Die Akademie von Moschopolis und ihre Nachwirkungen im Geistesleben Südosteuropas." In *Wissenschaftspolitik in Mittel- und Osteuropa: Wissenschaftliche Gesellschaften, Akademien und Hochschulen im 18. und beginnenden 19. Jahrhundert*, edited by Erik Amburger, Michał Cieśla, and László Sziklay, 114–28. Berlin: Ulrich Camen, 1976.

Peyfuss, Max Demeter. *Die Druckerei von Moschopolis, 1731–1769: Buchdruck und Heiligenverehrung im Erzbistum Achrida.* Vienna et al.: Böhlau, 1989.

Pink, Johanna. *Geschichte Ägyptens: Von der Spätantike bis zur Gegenwart.* Munich: C.H. Beck, 2014.

Plessner, Helmuth. *Die verspätete Nation.* Stuttgart et al.: Kohlhammer, 1959.

Pocock, John G.A. “Sprache und ihre Implikationen: Die Wende in der Erforschung des politischen Denkens.” In *Die Cambridge School der politischen Ideengeschichte*, edited by Martin Mulsow and Andreas Mahler, 88–126. Berlin: Suhrkamp, 2010.

Podskalsky, Gerhard. *Griechische Theologie in der Zeit der Türkenherrschaft (1453–1821): Die Orthodoxie im Spannungsfeld der nachreformatorischen Konfessionen des Westens.* Munich: C.H. Beck, 1988.

Polaschegg, Andrea. *Der andere Orientalismus: Regeln deutsch-morgenländischer Imagination im 19. Jahrhundert.* Berlin: De Gruyter 2003.

Polioudakis, Georgios. *Die Übersetzung deutscher Literatur ins Neugriechische vor der Griechischen Revolution von 1821.* Frankfurt/Main: Peter Lang, 2008.

Radtke, Bernd. “Erleuchtung und Aufklärung: Islamische Mystik und europäischer Rationalismus,” in *Die Welt des Islams*, Bd. 34 (1994): 48–66.

Radtke, Bernd. *Autochthone islamische Aufklärung im 18. Jahrhundert: theoretische und filologische Bemerkungen; Fortführung einer Debatte.* Utrecht: M.Th. Houtsma Stichting, 2000.

Raḍwān, Kamāl. *Almān fī Miṣr.* Cairo: al-Maktaba al-Qawmiyya al-Thaqafiyya, 1979.

Rāfi'ī, 'Abarraḥmān ar-. *Jamāladdīn al-Afghānī: bā'ith nahḍat ash-sharq, 1838–1897.* Cairo: Dar al-Ma'arif, 1991.

Rhonheimer, Martin. *Christentum und säkularer Staat: Geschichte – Gegenwart – Zukunft.* Freiburg, Basel and Vienna: Herder, 2012.

Rieber, Alfred. *The Struggle for the Eurasian Borderlands: From the Rise of Early Modern Empires to the End of the First World War.* Cambridge and New York: Cambridge University Press, 2014.

Riley-Smith, Jonathan. *The Crusades, Christianity, and Islam.* New York: Columbia University Press, 2008.

Riyāḍ, Umar. *Islamic Reformism and Christianity: A Critical Reading of the Works of Muhammad Rashid Rida and His Associates (1898–1935).* Leiden: Brill, 2008.

Rodrigue, Aron. “The Beginnings of Westernization and Community Reform Among Istanbul's Jewry, 1854–65.” In *The Jews of the Ottoman Empire*, edited by Avigdor Levy, 439–56. Princeton/New Jersey: The Darwin Press, 1994.

Rosenzweig, Franz. *Der Stern der Erlösung.* Heidelberg: Schneider, 1954.

Rouillard, Clarence D.. *The Turk in French History, Thought, and Literature (1520–1660).* Paris: Boivin, 1941.

Runciman, Steven. *Das Patriarchat von Konstantinopel: Com Vorabend der türkischen Eroberung bios zum griechischen Unabhängigkeitskrieg.* Munich: C.H. Beck 1970.

Rüsen, Jörn. *Kultur macht Sinn: Orientierung zwischen Gestern und Morgen.* Cologne et al.: Böhlau, 2006.

Said, Edward. *Orientalism.* New York: Pantheon, 1994.

Said, Edward. “Orientalism Reconsidered,” in *Race and Class* 27, Vol. II (1985): 1–15.

Said, Edward. *Culture and Imperialism.* London: Vintage 1994.

Ṣalībī, Kamāl aṣ-. “Islam and Syria in the Writings of Henri Lammens.” In *Historians of the Middle East*, edited by P.M. Holt, Bernard Lewis, and William M. Watt, 330–42. London: Oxford University Press, 1962.

Ṣalībī, Kamāl aṣ-. *tārīkh Lubnān al-ḥadīth.* Beirut: Dar an-Nahar, 1991.

Salvatore, Armando. *Islam and the Political Discourse of Modernity.* Reading: Ithaca Press, 1997.

Schluchter, Wolfgang. *Die Entstehung des modernen Rationalismus: Eine Analyse von Max Webers Entwicklungsgeschichte des Okzidents.* Frankfurt/Main: Suhrkamp, 1998.

Schmid, Wilhelm. *Die Geburt der Philosophie im Garten der Lüste.* Frankfurt/Main: Suhrkamp, 2000.

Schmidt-Biggemann, Wilhelm. "Political Theology in Renaissance Christian Kabbala: Petrus Galatinus and Guillaume Postel." In *Political Hebraism: Judaic Sources in Early Modern Political Thought*, edited by Gordon Schochet, Fania Oz-Salzberger and Meirav Jones, 3–28. Jerusalem and New York: Shalem Press, 2008.

Schmitt, Oliver Jens. *Levantiner: Lebenswelten und Identitäten einer ethnokonfessionellen Gruppe im osmanischen Reich im 'langen 19. Jahrhundert'.* Munich: R. Oldenbourg, 2005.

Schneiders, Werner. "Deus subjectum. Zur Entwicklung der Leibnizschen Metaphysik," Originally published in *Leibniz à Paris (1672–1676)*, Vol. II, *La philosophie de Leibniz, Studia Leibnitiana, Supplementa*, edited by K. Müller, H. Schepers and W. Totok, Vol. XVIII, Wiesbaden: Steiner 1978: 21–31, repr. in *Werner Schneiders. Philosophie der Aufklärung – Aufklärung der Philosophie: Gesammelte Studien, zu seinem 70. Geburtstag*, edited by Frank Grunert, 13–24. Berlin: Duncker & Humblot, 2005.

Schneiders, Werner. "Harmonia universalis. Harmonie als Schlüsselbegriff der Leibnizschen Philosophie," originally published in *Studia Leibnitiana*, Vol. XVI/1 (1984): 27–44, reprinted in *Philosophie der Aufklärung – Aufklärung der Philosophie: Gesammelte Studien, zu seinem 70. Geburtstag*, edited by Frank Grunert, 25–47. Berlin: Duncker & Humblot, 2005.

Schölch, Alexander. "Der arabische Osten im neunzehnten Jahrhundert, 1800–1914." In *Geschichte der arabischen Welt*, edited by Ulrich Haarmann and Heinz Halm, 365–431. Munich: C.H. Beck, 2004.

Schubart, Wilhelm. *Justinian und Theodora.* Munich: F. Bruckmann, 1943.

Schulze, Reinhard. "Das islamische achtzehnte Jahrhundert. Versuch einer historiographischen Kritik," in *Die Welt des Islams* 30 (1990): 140–59.

Schulze, Reinhard. "Was ist die islamische Aufklärung?" in *Die Welt des Islams* 36 (1996): 276–325.

Schulze, Reinhard. "Gibt es eine islamische Moderne?," in *Der Islam und der Westen: Anstiftung zum Dialog*, edited by Kai Hafez, 31–43. Frankfurt/Main: Fischer 1997.

Schulze, Reinhard. "Islam und Judentum im Angesicht der Protestantisierung der Religionen im 19. Jahrhundert." In *Judaism, Christianity, and Islam in the Course of History: Exchange and Conflicts*, edited by Lothar Gall and Dietmar Willoweit, 139–65. Munich: R. Oldenbourg, 2011.

Seikaly, Samir M. "Damascene Intellectual Life in the Opening Years of the 20th Century: Muhammad Kurd 'Ali and *Al-Muqtabas*." In *Intellectual Life in the Arab East, 1890–1939*, edited by Marwan R. Buheiry, 125–53. Beirut: American University of Beirut, 1981.

Scott, Hamish. *The Birth of a Great Power System, 1740–1815.* Harlow and New York: Pearson/Longman, 2006.

Shagrir, Iris, and Nitzan Amitai-Preiss. "Michaud, Montrond, Mazloum and the First History of the Crusades in Arabic," in *Al-Masāq*, Vol. 24, No. 3, December 2012: 309–12.

Sharabi, Hisham. *Arab Intellectuals and the West: The Formative Years, 1875–1914.* Baltimore: John Hopkins Press, 1970.

Sharabi, Hisham. *al-muthaqqafūn al-ʿarab wa-l-gharb, ʿaṣr an-nahḍa 1875–1914* [Arabic transl. of *Arab intellectuals*]. Beirut: Dar an-Nahar, 1991.
Sloterdijk, Peter. *Gottes Eifer: Vom Kampf der drei Monotheismen*. Frankfurt/Main and Leipzig: Verlag der Weltreligionen, 2007.
Spengler, Oswald. *Der Untergang des Abendlandes: Umrisse einer Morphologie der Weltgeschichte*. Munich: C.H. Beck, 1990.
Stauth, Georg, and Marcus Otto. *Méditerranée: Skizzen zu Mittelmeer, Islam und die Theorie der Moderne*. Berlin: Kadmos, 2008.
Stone, Lawrence. "Der Wandel der Werte in England 1660 bis 1770: Säkularismus, Rationalismus und Individualismus." In *Kulturen der Achsenzeit*, Bd. II: *Ihre institutionelle und kulturelle Dynamik*, Teil 3: *Buddhismus, Islam, Altägypten, westliche Kultur*, edited by Shmuel N. Eisenstadt, 341–57. Frankfurt/Main: Suhrkamp, 1992.
Straßenberger, Grit. "Hannah Arendt, Michael Walzer und Martha Craven Nussbaum." In *Politische Ideengeschichte im 20. Jahrhundert: Konzepte und Kritik*, edited by Harald Bluhm and Jürgen Gebhardt, 155–80. Baden-Baden: Nomos, 2006.
Strohmaier, Gotthard. "Ethical Sentences and Anecdotes of Greek Philosophes in Arabic Tradition." In *Von Demokrit bis Dante*. Hildesheim et al.: Georg Olms, 1996: 44–52.
Strohmaier, Gotthard. *Avicenna*. Munich: C.H. Beck, 1999.
Strohmaier, Gotthard. "Was Europa dem Islam verdankt." In *Hellas im Islam: Interdisziplinäre Studien zur Ikonographie, Wissenschaft und Religionsgeschichte*. Wiesbaden: Harrassowitz, 2003: 1–27.
Sundhaussen, Holm. *Der Einfluss der Herderschen Ideen auf die Nationsbildung bei den Völkern der Habsburger Monarchie*. Munich: R. Oldenbourg, 1973.
Suppé, Frank-Thomas. "In Sachsen auf Heimatboden. Zur Geschichte der griechischen Gemeinde in Leipzig von ihren Anfängen bis nach 1945." In *Eugênios Búlgaris und die griechische Aufklärung in Leipzig: Die Griechen im Leipzig des 18. Jahrhunderts; Eine Ausstellung der Universitätsbibliothek Leipzig vom 16. Bis 30. Oktober 1996*, 13–48. Leipzig: Universitätsbibliothek 1996.
Tamcke, Martin. *Das orthodoxe Christentum*. Munich: C.H. Beck, 2004.
Taraman, Soheir. *Kulturspezifik als Übersetzungsproblem: Phraseologismen in arabisch-deutscher Übersetzung*. Heidelberg: Julius Groos, 1986.
Tavakoli-Targhi, Mohamed. *Refashioning Iran: Orientalism, Occidentalism and Historiography*. Basingstoke and New: Palgrave York, 2001.
Tcherikover, Victor. *Hellenistic Civilization and the Jews*. Peabody/Mass.: Hendrickson, 1999.
Thielmann, Jörn. *Naṣr Ḥāmid Abū Zaid und die wiedererfundene ḥisba: Sharīʿa und Qānūn im heutigen Ägypten*. Würzburg: Ergon, 2003.
Tibawi, Abdul Latif. "The Genesis and Early History of the Syrian Protestant College." In *American University of Beirut Festival Book* (Festschrift), edited by Fuad Sarruf and Suha Tamim, 257–94. Beirut: Eastern Printing, 1967.
Tibi, Bassam. *Der Islam und das Problem der kulturellen Bewältigung sozialen Wandels*. Frankfurt/ Main: Suhrkamp, 1985.
Tibi, Bassam. *Vom Gottesreich zum Nationalstaat. Islam und panarabischer Nationalismus*. Frankfurt/ Main: Suhrkamp, 1987.
Tibi, Bassam. *Islamischer Fundamentalismus, moderne Wissenschaft und Technologie*. Frankfurt/Main: Suhrkamp, 1992.

Todorova, Maria N. "Historische Vermächtnisse als Analysekategorie. Der Fall Südosteuropa." In *Wieser Enzyklopädie des Europäischen Ostens, Vol. 11: Europa und die Grenzen im Kopf*, edited by Karl Kaser, Dagmar Gramshammer-Kohl and Robert Pichler, 227–52. Klagenfurt, Vienna and Ljubljana: Wieser, 2003.

Toner, Jerry. *Homer's Turk: How Classics Shaped Ideas of the Middle East.* Cambridge/Mass. and London: Harvard University Press, 2013.

Tönnies, Sibylle. *Der westliche Universalismus: Eine Verteidigung klassischer Positionen.* Opladen: Westdeutscher Verlag, 1995.

Traut, Tobias. "Der Staat im Denken der Russisch-Orthodoxen Kirche: Platz für Demokratie?" In *Religion in Diktatur und Demokratie: Zur Bedeutung religiöser Werte, Praktiken und Institutionen in politischen Transformationsprozessen*, edited by Simon Wolfgang Fuchs and Stephanie Garling, 59–78. Münster and Berlin: Lit, 2011.

Troeltsch, Ernst. *Der Historismus und seine Probleme. Erstes (einziges) Buch: Das logische Problem der Geschichtsphilosophie.* Aalen: Scientia Verlag, [2nd repr. of the ed. Tübingen 1922] 1977.

Turan, Ömer. "American Protestant Missionaries and Monastir, 1912–17: Secondary Actors in the Construction of Balkan Identities," in *Middle Eastern Studies*, Vol. 36, No. 4, Oct. 2000: 119–36.

Turczynski, Emanuel. "Gestaltwandel und Trägerschichten der Aufklärung in Ost- und Südosteuropa." In *Die Aufklärung in Ost- und Südosteuropa: Aufsätze, Vorträge, Diskussionen*, edited by Erna Lesky, Strahinja K. Kostić, Josef Matl and Georg von Rauch, 23–49. Cologne and Vienna: Böhlau, 1972.

Ucko, Peter J., and T. C. Champion. *The Wisdom of Egypt: Changing Visions Through the Ages.* London: UCL Press, 2003.

Urvoy, Dominique. *Ibn Rushd (Averroes).* London and New York: Routledge, 1991.

Vakalopoulos, Apostolos. *Griechische Geschichte von 1204 bis heute.* Cologne: Romiosini, 1985.

Vacalopoulos [Vakalopoulos], Apostolos. "Byzantinism and Hellenism," in *Balkan Studies*, Vol. 9 (1968): 101–26.

Watenpaugh, Keith D. *Being Modern in the Middle East: Revolution, Nationalism, Colonialism, and the Arab Middle Class.* Princeton/New Jersey: Princeton University Press, 2006.

Weber, Christina. *Die jüdische Gemeinde im Damaskus des 19. Jahrhunderts: Städtische Sozialgeschichte und osmanische Gerichtsbarkeit im Spiegel islamischer und jüdischer Quellen.* Berlin: Klaus Schwarz, 2011.

Weingrod, Alex. "Saints and shrines, politics and culture: a Morocco–Israel comparison." In *Muslim Travellers: Pilgrimage, Migration, and the Religious Imagination*, edited by Dale F. Eickelman and James Piscatori, 217–35. London: Routledge 1990.

Wiggers, Julius. *Geschichte der Evangelischen Mission.* 2 vols., Hamburg and Gotha: Perthes, 1845.

Wilsdorf, Helmut. "Georgius Agricola, die 'jüngeren Griechen' und das Morgenland." In *Griechenland – Byzanz – Europa: ein Studienband*, edited by Joachim Herrmann, Helga Köpstein and Reimar Müller, 215–24. Amsterdam: Gieben, 1988.

Winter, Jakob, and August Wünsche. *Geschichte der jüdisch-hellenistischen und talmudischen Litteratur: Zugleich eine Anthologie für Schule und Haus.* Vol. 1: *Die jüdische Litteratur seit Abschluss des Kanons.* Trier: S. Mayer, 1894.

Woodhouse, Christopher. *Rhigas Velestinlis: The Proto-Martyr of the Greek Revolution.* Limni Evia: D. Harvey, 1995.
Wokart, Norbert. *Kontaminationen: Antike Spuren in unserem Denken.* Würzburg: Königshausen und Neumann, 2014.
Xydis, Stephen G. "Modern Greek Nationalism." In *Nationalism in Eastern Europe*, edited by Peter F. Sugar and Ivo John Lederer, 207–58. Seattle: University of Washington Press, 1994.
Zahlan, Antoine B. "The Impact of Technology Change on the Nineteenth-Century Arab World." In *Between the State and Islam*, edited by Charles E. Butterworth and I. William Zartman, 31–58. Cambridge and Washington: Woodrow Wilson Center Press, 2001.
Zanou, Konstantina. "Nostalgia, Self-Exile and the National Idea: The Case of Andrea Mustoxidi and the Early Nineteenth-Century Heptanesians of Italy." In *Nationalism in the Troubled Triangle: Cyprus, Greece and Turkey*, edited by Ayhan Aktar, Niyazi Kızılyürek and Umut Özkırımlı, 98–111. London et al.: Palgrave Macmillan 2010.
Zelepos, Ioannis. *Die Ethnisierung griechischer Identität 1870–1912: Staat und private Akteure vor dem Hintergrund der ‚Megali Idea'.* Munich: R. Oldenbourg: 2002.
Ziedan, Youssef [Yūsif Zīdān]. *al-lāhūt al-ʿarabī wa-uṣūl al-ʿunf ad-dīnī.* Cairo: Dar ash-Shuruq, 2010.

Index

ʿAbbās I (Shāh) 13, 160
ʿAbbās (khedive) 65
ʿAbdarrāziq, ʿAlī 36, 73, 164
ʿAbduh, Muḥammad 36, 69, 73, 137-8
ʿAbdülḥamīd II (sultan) 70, 132, 163
Abū Ḥanīfa, an-Nuʿmān b. Thābit 29, 32-3
Abū Qurra, Theodore 36
Abū Zayd, Naṣr Ḥāmid 29–34, 156
Ṣābūnjī, Louis aṣ- 145
Adham, Ismāʿīl Aḥmad 86–89, 109–111, 127-8
Adorno, Theodor W. 24
Afghānī, Jamāladdīn al- 69, 73, 106
Agioritis, Nikodimos 49
Ṭahṭāwī, Rifāʿa Rāfiʿ aṭ- 111-2, 162
Ḥakīm, Tawfīq al- 86f., 164
Alekseevich, Fedor (tsar) 40
Alexander the Great 52-3, 56, 116-7
Ali Pasha of Iannina 59, 75
Amīn, Qāsim 73, 79–85, 92, 101-2, 106, 164
ʿAmūn, Iskandar Bey 145
Amun (ancient god) 116
Anthimos of Jerusalem (patriarch) 49, 147
Anthrakitis, Methodios 41, 161
Anṭūn, Faraḥ 71, 137-8, 145, 164
Argentis, Eustratios 39, 41
Argyropoulos, Loukas 75
Aristotle 134
Ḥassūn, Rizqallāh 145
Aswānī, ʿAlāʾ al- 128
Atatürk, i.e. Mustafa Kemal 108-9
Averroes (i.e. Ibn Rushd) 137, 155

Bacon, Francis 23, 93, 139
Bacon, Roger 90
Baghdādī, Abū l-Barakāt al- 35
Bannā, Ḥasan al- 147
Bardas 27
Barqāwī, Aḥmad Nasīm 17, 35-6, 66-7, 69, 72, 101, 126, 144–146
Barrès, Maurice 114, 150
Bārshīyā, Ibrāhīm 94
Bauer, Thomas 20
Berlin, Isaiah 24-5
Bilharz, Theodor 66
Blumenberg, Hans 22, 24-5, 51, 54, 93, 157
Bretonne, Nicolas Restif de la 57
Brockelmann, Carl 58
Bubner, Rüdiger 24
Buchenau, Klaus 84
Burke, Edmund 11, 14
Bustānī, Sulaymān al- 72, 99-100, 103–105, 129–132, 164

Carlyle, Thomas 22, 163
Cassirer, Ernst 21, 24
Catherine II (Russian empress) 14, 44-5, 55, 120, 161
Catherine the Great 14, 44-5, 55, 120, 161
Chamberlain, Neville 147
Charles V (king) 55
Chevalier, Michel 64, 162
Chiha, Michel 150
Clark, Gregory 49
Clot Bey (i.e. Bartholomew Clot) 63
Cole, Juan R. 53, 78
Comte, Auguste 87-8
Condorcet, Marie-Jean-Antoine-Nicolas de Caritat 25
Cook, Michael 26
Cosmas of Aetolia 77
Crusius, Martin 13, 160
Cudworth, Ralph 64
Cvijić, Jovan 99

Da Vinci, Leonardo 92
Damodos, Vikentios 41, 43-4
David, Julius 48
Descartes, René 21, 25, 41, 48, 66, 139
Desiderius Erasmus 92, 139
d'Herbelot, Barthélemi 160
Diderot 93
Diner, Dan 68
Doukas, Neofitos 48, 62

https://doi.org/10.1515/9783110634006-010

Dragoumis, Ionos 57, 60–62, 77, 98–100, 120–122, 128, 134–136, 149–153, 164

Elcalay, Mercado 72
Eparchos, Antonios 13

Fallmerayer, Jacob Philipp 57, 119
Fārābī, Abū Naṣr Muḥammad 35, 111, 155
Fares, Felix 127-8
Farmakidis, Theoklitos 62, 76
Flasch, Kurt 25
Florovsky, Georges 149
Fotiadis, Lambros 48
Frederick II (king) 96
Fukuyama, Francis 20, 26, 75, 153, 158

Galland, Antoine 14
Gama, Vasco da 114
Gazis, Anthimos 59
Gennadios II (patriarch) 132
Germanos (archbishop) 59
Ghalioun, Burhan 141–143, 154-5
Ghazālī, Abū Ḥāmid Muḥammad al- 155
Gibb, H.A.R. 95
Gibbon, Edward 16
Gobineau, Arthur 89
Goethe, Johann Wolfgang von 126
Gökalp, Ziya 84, 106, 108-9
Goldziher, Ignaz 35
Gregory V (patriarch) 53, 59, 125
Griesinger, Wilhelm 66
Grunebaum, Gustav E. von 34, 100
Guilford, Frederick North 5th Earl of 59, 162

Haffner, Sebastian 12
Haga, Cornelis 38
Hagen, Gottfried 18
Hartmann, Martin 78
Hatzidakis, Georgios 15
Haykal, Muḥammad Ḥusayn 112
Helvétius, Claude-Adrien 25, 46
Herder, Johann Gottfried 14, 23, 45, 134, 45, 134
Herman, Arthur 22
Hilarion (archimandrite) 125, 162
Himmelfarb, Gertrude 23
Hitler, Adolf 147
Hobbes, Thomas 46
Holbach, Paul-Henri Thiry (Baron) 25
Homer 127-8
Horkheimer, Max 24
Huntington, Samuel P. 12

Ibn ʿĀbidīn, Muḥammad Amīn b. ʿUmar 81
Ibn Ezra 94
Ibn Rushd 94, 111, 137
Ibn Sīnā 94, 111
Ibrāhīm b. Ḥiyā 94
Ibrāhīm (Egyptian khedive) 61
Idrīsī, Abū ʿAbdallāh 114
Isḥāq, Adīb 145
Ismāʿīl (khedive) 66
Iustinian (emperor) 74

Jefferson, Thomas 62
Jisr aṭ-Ṭarābulusī, Ḥusayn al- 70-1, 86
John of Damascus 28, 36

Kairis, Theofilos 62
Kallioupolitis, Maximos 123
Kant, Immanuel 25, 42, 46-7
Kapodistrias, Ioannis 59
Karam, Yūsif Beg 130, 163
Karsh, Efraim 147
Katartzis, Dimitrios 55, 62
Kavalliotis, Theodoros Anastasios 47-8
Kawākibī, ʿAbdarraḥmān al- 85
Khwārizmī, Muḥammad b. Mūsā al- 94
Klat, Hector 150
Kohn, Hans 52
Kolettis, Ioannis 59
Komitas, Stefanos 48, 62
Korais, Adamantios 40, 46, 61-2, 124
Koridalleas, Theofilos 38, 160
Koumas, Konstantinos 42
Kozlarek, Oliver 20
Kritopoulos, Mitrofanis 38
Krug, Wilhelm Traugott 42
Krumbacher, Karl 15, 164
Kurd ʿAlī, Muḥammad 107
Kyrillos V (patriarch) 41, 161

Lammens, Henri 114
Lapidus, Ira M. 27, 34

Lazarus-Yafeh, Hava 155
Le Bon, Gustave 137
Leibniz, Gottfried Wilhelm 43-4, 46, 48, 51
Leontiadis, Sevastos 48
Lewis, Bernard 15, 78, 112, 158
Ligaridis, Paisios 40
Locke, John 23, 43-4, 46
Lord Byron (George Gordon) 120, 162
Louis XIV (king) 51
Loukaris, Kyrillos (patriarch) 38, 123, 160
Luther, Martin 11, 19, 36, 69, 90, 92, 139

Maʿarrī, Abū al-ʿAlāʾ al- 35, 97
Machiavelli, Niccolò 140
MacPherson, James 134
Maimonides 91
Makarios (metropolitan of Corinth) 49
Makrides, Vasilios N. 148, 153
Malebranche, Nicolas de 41, 48
Maḥmūd, Zakī Najīb 115
Maḥmūd II (Ottoman sultan) 58, 66
Margounios, Maximos 13, 123
Marrāsh, Francis Fatḥallāh al- 71, 145, 163
Martel, Charles 89
Maurer, Georg Ludwig von 55, 59–62, 74–76, 133, 148, 153, 162
Mavrokordatos, Alexandros 60
Mavrokordatos, Nikolaos 123
Maximos Maẓlūm III (Melkite patriarch) 144
Maximos of Gallipoli 38, 123
Melanchthon, Philipp 13
Michaud, Joseph-François 144, 162
Midḥat Pasha 71-2, 163
Miṣrī, ʿAzīz al- 145
Mohammed (prophet) 52
Montesquieu, Charles-Louis de Secondat 23, 50, 56, 66
Moschopoulos, Antonios 42f., 161
Muḥammad ʿAlī (viceroy) 61, 63–67, 69, 88, 120, 162-3
Muḥammad b. ʿAbdallāh (prophet) 29-30, 36, 51, 73, 88, 138, 140
Mudawwar, Michel 145
Mūsā, Salāma 71, 90–97, 105, 108 f., 112, 115 f., 127, 138 f., 141, 147, 164
Mūsā b. Ṭibbūn 94

Nagel, Tilman 28f.
Napoleon, Bonaparte 51–54, 65, 105
Napoleon Bonaparte 49–54, 59, 65, 102, 105, 161-2
Napoleon III (ruler) 68
Nelson, Horatio (admiral) 54
Newton, Isaac 23, 93
Nicolaus I (tsar) 68
Nietzsche, Friedrich 53, 127
Nikolaou, Theodor 40
Norden, Frederik 50

O'Donnell, Kathleen Ann 133
Oekonomos brothers 62
Ossian 126, 133-4
Osterhammel, Jürgen 13, 158
Otto of Bavaria (king) 74, 148
Ovid 127

Palaiologos, Michalis 120
Palamas, Gregory 27, 37, 40, 148, 160
Pamplekis, Christodoulos Eustathiou 47, 161
Paraskevas, Damianos 42-3
Parezanin, Ratko 110
Parios, Athanasios 48-9, 148, 161
Paschidis, Thomas 134
Pegas, Meletios 13, 123
Pericles 122
Peter Abelard 91
Peter the Great 55
Pindar 127
Plato 134, 141
Plessner, Helmuth 19
Plethon, Georgios Gemistos 37
Plutarch 122
Pococke, Richard 50
Podskalsky, Gerhard 40, 43, 46
Polo, Marco 114
Polyeidis, Theokletos 39, 161
Postel, Guillaume 15
Prokopios, Dimitrios 124
Prokopios (monk) 148
Psalidas, Athanasios 46-7

Qudsī, Ḥāfiẓaddīn al- 50

Radtke, Bernd 18
Reclus, Elisée 114
Renan, Ernest 80, 137
Rhonheimer, Martin 25
Rhyosis, Diamantis 40
Riḍā, Rashīd 69
Riggs, Elias 126, 163
Riley-Smith, Jonathan 144
Roger II (ruler of Sicily) 114
Rosenzweig, Franz 26
Rousseau, Jean-Jacques 24, 47, 50, 56, 66, 93, 134

Said, Edward 158
Sa'īd (khedive) 66
Saladdin 91, 120
Salvatore, Armando 19
Schlözer, August Ludwig 23, 45
Scholarios, Georgios i.e. Gennadios II (patriarch) 132
Schulze, Reinhard 17–19
Scott, Walter 143
Seidensticker, Tilman 18
Selim III (sultan) 48-9
Semler, Johann Salomo 64
Shāfi'ī, Muḥammad b. Idrīs ash- 29–34, 81
Shahbandar, 'Abdarraḥmān 140
Shāhīn, Ṭanyūs 130, 163
Sharābī, Hishām 54
Sharqāwī, 'Abdullāh ash- 52
Shidyāq, Aḥmad Fāris ash- 54, 83, 163
Shumayyil, Shiblī ash- 132
Skoufas, Nikolaos 58
Sloterdijk, Peter 24-5
Socrates 122
Sofianos, Sofianos 122
Sokrates 21
Sophia-Louise (queen) 124
Sophocles 122
Spencer, John 64
Spengler, Oswald 149
Spinoza, Baruch 46-7
Steinberg, Gerald M. 157
Süleyman I (sultan) 55

Talibov, 'Abdarraḥmān 107
Themistocles 122
Theophilos of Kampania (bishop) 148
Theotokis, Nikiforos 47
Theseus 150
Thomas Aquinas 91
Thomasius, Christian 43
Thunmann, Johann 48
Todorova, Maria 99
Toner, Jerry 4, 16
Tott, Baron de 67
Troeltsch, Ernst 111
Tsakaloff, Athanasios 58

'Uraysī, 'Abdalghanī al- 145
Ḥuṣrī, Ṣāṭi' al- 146
Ḥusayn, Ṭāhā 35, 97, 112, 116–118, 140-1, 147, 164
Ḥusayn b. 'Alī (sherif) 147
Ḥusaynī, Ḥājj Amīn al- 147

Vamvas, Neofitos 62
Velestinlis, Rigas Ferraios 55–57, 161
Velimirovic, Nikolay 152
Voltaire, François Marie Arouet de 25, 42–44, 46-7, 54, 56, 66, 93, 120
Voulgaris, Eugenios 41–48, 77, 123, 148, 161

Weber, Max 24
Werry, Francis 133
Wilson, Woodrow 20
Wolff, Christian 42–44, 46, 48, 161

Xanthos, Emmanouil 58

Ypsilantis, Alexandros 120

Zambelios, Spyridion 119
Zanjānī, Shaykh Ebrāhīm 85
Zaqzūq, Muḥammad Ḥamdī 29, 128
Zaydān, Jurjī 83-4, 112, 132, 164
Ziedan, Youssef 155
Zīla'ī, 'Uthmān b. 'Alī az- 81
Zygabenos, Euthymios 37
Zygomalas, Theodosios 13

www.ingramcontent.com/pod-product-compliance
Lightning Source LLC
Chambersburg PA
CBHW032100230426
43662CB00035B/857
* 9 7 8 3 1 1 0 6 4 3 5 6 5 *